Community Eldercare Ecology in China

Lin Chen · Minzhi Ye

Community Eldercare Ecology in China

palgrave
macmillan

Lin Chen
Fudan University
Shanghai, China

Minzhi Ye
Case Western Reserve University
Cleveland, OH, USA

ISBN 978-981-15-4959-5 ISBN 978-981-15-4960-1 (eBook)
https://doi.org/10.1007/978-981-15-4960-1

This Palgrave Macmillan imprint is published by the registered company Springer Nature Singapore Pte Ltd.
The registered company address is: 152 Beach Road, #21-01/04 Gateway East, Singapore 189721, Singapore

Our parents

Acknowledgments

Lin:

This book is among many products from the collaboration between me and Minzhi in the past decade. We met at the annual meeting of the Gerontological Society of America in New Orleans in 2010. I stopped in front of her poster on the community-based eldercare in Shanghai and we began talking. As we talked more, I found out that both of us are from Shanghai and interested in aging, community, and long-term care. At that time, I just started my third year doctoral training in social welfare at UCLA and she was a masters student at Bowling Green. Since then, we have published 6 peer-reviewed journal articles together, collaborated on 2 research projects, and this book. Ten years later, I am now an Associate Professor in Social Work at Fudan University and she is going to be garnered with a Ph.D. in Sociology from Case Western Reserve University.

This journey has formed most of my research agenda and academic pursuit for knowing about older adults, their needs for care, and their life in the community. I am grateful to so many seasoned scholars, my doctoral mentor, Dr. Lené Levy-Storms; my doctoral committee members: Professor Laura Abrams, Professor Ted Benjamin, and Prof. Marjorie Kagawa-Singer from UCLA. Professor Iris Chi from Social Work at USC has been my GSA mentor since 2011. Dean and Professor Sarah Gehlert from USC Suzanne Dworak-Peck School of Social Work, then Dean and Professor from College of Social Work at University of South Carolina; Professor Wen-Jui Han from NYU; Professor Sue Levkoff from University of South Carolina; Professor Dale Dannefer from Case Western Reserve University; Professor Shenyang Guo, Professor Carolyn Lesorogol, and Ms. Linyun Fu from Washington University in St. Louis cordially offered me opportunities to share some early versions of ideas and chapters of this book during my visiting in 2018. Professor Eva Kahana from Case Western Reserve University deserves my deep appreciation. Professor Kahana is Minzhi's doctoral mentor while she has also graciously taken me under her wings since 2014. We cannot have this book or several journal articles without Professor Kahana's generous support.

Ms. Nancy Lee Sayre, Thank you. You are the hidden figure of this book and almost all my publications.

My colleague, Felicia Tian, thank you for nudging me to write the proposal for this book.

And finally, my mother, happy 70th birthday this October.

Minzhi:

Writing a book is a challenging yet rewarding and meaningful journey. It would not have been possible without my guide and co-author, Dr. Lin Chen. She understood my every struggle and provided invaluable help. I'm eternally grateful to my parents, Luhua Ye and Fengchao Luo. I could not have reached my goal without their most powerful and loving support.

I would also like to express my gratitude to my mentor, Dr. Eva Kahana. She provided essential advice on the topic selection and chapter arrangement of the book. Many friends generously read and intelligently commented on the early drafts of this book. They are James Stephens, Micheal Slone, Camille Lartaud-Balosso, Henrique Rodrigues, George Burke, Sophie Shi, Rock Lim, Anna Fedotova, and Siddhesh Ambhire. I also want to thank all my friends, who encouraged me along the way. Without their unstinting care and support, it would have been difficult for me to complete the manuscript. Finally, I want to thank the Steiner house for providing a warm home and the necessary resources for me to finish the project.

Lin & Minzhi:

This book is in debt to many scholars and practitioners in Shanghai: Mr. Pengbiao Sun, Ms. Xiaohua Li, Mr. Yuebin Chen, Dr. Xuejin Zuo, Dr. Qingxuan Liu, Mr. Lin Gu, Ms. Yiwen Chen. Your experiences of and devotion to taking care of older adults in Shanghai has tremendously moved and motivated us. We also want to express our gratitude for all the older adults and staff members of the HCBS program in the Jing'an District.

We thank Palgrave Macmillan, especially Sara Crowley Vigneau and Connie Li, for all the publication support.

Finally, we thank SAGE publication for permission to use extracts of our analysis on the early stage of this project:

1. Chen, L., Ye, M., & Kahana, E. (2018). "Their today is our future": Direct carers' work experience and formal caring relationships in a community-based eldercare program in Shanghai. *Journal of Applied Gerontology*, 37, 516–537. doi: 10.1177/0733464816653360.
2. Ye, M., Chen, L., & Kahana, E. (2017). Mealtime interactions and life satisfaction among older adults in Shanghai. *Journal of Aging and Health*, 29(4), 620–639. doi: 10.1177/0898264316641080.

Contents

List of Figures

List of Tables

CHAPTER 1

Introduction

Facing potentially high demands for long-term care of its fast-growing aging population, Shanghai has been devoted to building a sustainable long-term care system since 2001 (Chen & Han, 2016). Two decades ago, the Shanghai government began to develop and implement home- and community-based services (HCBS) for the increasing number of older adults in the city. Wu, Carter, Goins, and Cheng (2005) reviewed Shanghai's community-based long-term care system in the early 2000s, analyzing its benefits for community-dwelling older adults and praising its emergence. Chen and Han (2016) reviewed the most recent developments in community-based eldercare policy in Shanghai between 2007 and 2015. Our book builds upon this literature by focusing on the recent development of an HCBS program in the Jing'an District in Shanghai, which has emerged since 2001 and begun to fully develop since 2007. We chronicle its progress, along with the overall development of HCBS in Shanghai.

Indeed, HCBS was a new concept for Chinese older adults. Because eldercare in China has historically relied on family members, especially adult children, introducing HCBS has been an innovative way to share caregiving responsibilities across family and community. Some people have welcomed the idea while others have resisted it; Chinese families have witnessed a trend of evolving caregiving: adult children are sometimes invested in the tradition of taking care of their older parents at home

L. Chen and M. Ye, *Community Eldercare Ecology in China*,
https://doi.org/10.1007/978-981-15-4960-1_1

and sometimes choose other caregiving alternatives instead. Scholars, policymakers, and even laypeople have been heated debates on this subject. In light of the traditional expectations, does this generation of Chinese adult children really want to relinquish family caregiving responsibility by taking advantage of HCBS? However, some older adults have welcomed HCBS, first because they did not have to prepare their meals every day and second, because they were able to spend time in the community and make new friends. Based on feedback from various stakeholders and especially service recipients—older adults themselves—the Shanghai Civil Affairs Bureau has published and implemented a series of policies, regulations, and standards to develop HCBS for older adults living in the community (Table 1.1).

By reviewing one of the most advanced HCBS programs in Shanghai, now 13 years old, this book aims to provide fresh insight into community care ecology for eldercare in urban China. Taking a social-ecological perspective, we view the community as an organic system, where older adults interact on the macro, meso, and micro levels of the community care ecology. Drawing on mixed-method approaches, including surveys, in-depth interviews, and government archives, we explore the emergence of HCBS in Shanghai, its development over the past decade, its administration and services offered, its resource allocation, staff members' work experiences, older adults' service experiences, and service evaluation and improvements. Both quantitative and qualitative data illuminate multilayered interactions among these aspects of the community ecology. The purpose of our analysis is to show how the HCBS program in the Jing'an District has addressed older adults' needs, influenced their interpersonal and social networking dynamics, fostered a new form of caregiving relationship in the community setting, and shaped a new discourse on caregiving policy. Our theoretical analysis reflects the implications of the changing sociocultural context for eldercare in urban China.

1.1 Community Ecology and Aging

Gerontologists have long viewed community as a perfect venue for aging and have proposed the concept of "aging in place," which refers to "remaining living in the community, with some level of independence, rather than in residential care" (Davey, Nana, de Joux, & Arcus, 2004, p. 133). While growing research shows the benefits of aging in place for older adults, some scholars have pointed out that the essential element

Table 1.1 Chronology of the policy development of community-based eldercare published by Shanghai Civil Affairs Bureau since 2001

Date	*Policy*
April 10, 2001	Opinions on the comprehensive development of home-based care services: "To adapt to the increasing aging population in Shanghai, home- and community-based services for older adults should be promoted to the entire city."
November 4, 2003	Notice on further deepening the pilot work of home- and community-based eldercare services
April 20, 2004	Notice on further promoting and deepening home- and community-based eldercare services: "The Shanghai government plans to establish a comparatively comprehensive home- and community-based eldercare service network with good operation mechanisms in the next two years."
October 20, 2006	Opinions on further promoting the development of eldercare services in Shanghai: "In the eleventh Five-Year Plan... home- and community-based eldercare services in Shanghai should cover 250 thousand older adults, about 8% of the total aging population."
April 2, 2008	Notice on encouraging communities to set up community cafeteria for older adults: "In 2008, there will be 200 service centers for older adults to dine or take out in the community across the city. Each service center should provide for at least 50 people for every meal."
June 9, 2009	Notice on further regulating home- and community-based eldercare services in Shanghai
February 11, 2010	Notice on implementing Shanghai local standards on home- and community-based eldercare services
March 19, 2014	Opinions on adjusting relevant policies on home- and community-based eldercare services in Shanghai
December 11, 2019	Notice on distributing the "Guidelines for embedded home- and community-based eldercare services in Shanghai"

of aging in place is to build an engaging, inclusive, and interdependent environment to offer sustainable, healthy opportunities for older adults (Thomas & Blanchard, 2009)—"an adaptive process of ongoing person-place transactions over time" (Scharlach & Moore, 2016, p. 420). In a recent review of the evolving topics and studies in the field of aging in place, scholars have investigated shifts in research focus from housing and environmental modification to interpersonal interactions, care, and services (Vasunilashorn, Steinman, Liebig, & Pynoos, 2012).

More recently, scholars have further distinguished "aging in community" from "aging in place" (e.g., Blanchard, 2013; Greenfield, Black, Buffel, & Yeh, 2019); aging in place indicates a relatively static status, such as remaining in an apartment or other housing, whereas aging in community suggests older adults' spatial, geographical, and interpersonal relationships with the community where they live, as well as with various organizations and sectors of the community (Blanchard, 2013).

Although the concept of community has remained elusive and kept evolving since its emergence in the late nineteenth century (e.g., Ahn, Kwon, & Kang, 2020; Chipuer & Pretty, 1999; Putnam, 2000), in this book, we follow the notion that community provides geographic proximity, social networks, feelings of belonging, services, and so forth, to support older adults' daily life and further sustain their independence and autonomy (e.g., Callahan, 1993; Chen, Ye, & Kahana, 2019; Lawler, 2001; Palley, 2009; Wiles, 2005). In fact, community covers all aspects, arrangements, and organizations in individuals' aging process; that is, their aging process is intertwined with their life in the community (Greenfield et al., 2019). Both the World Health Organization (WHO) and the (American) Centers for Disease Control and Prevention (CDC) have recommended that older adults remain in their familiar communities to benefit their aging process, including their physical and psychological well-being (CDC, 2013; WHO, 2007). Aging in community thus becomes an affordable alternative to institutional caregiving, reducing high instrumental, financial, and psychological costs for older adults and their family caregivers, and gaining favor from both policymakers and healthcare service experts (WHO, 2007).

Reflecting the distinction between aging in place and aging in community, research focuses on this topic have also changed from the static physical environment to the more dynamic relationship between older adults and social environment in the community. Buffel et al. (2012) further theorized three aspects of the relationship between older adults and community: geographical neighborhoods (i.e., housing and local attachment), social environments (i.e., interpersonal relationships and sense of security), and social engagement (i.e., political advocacy, policy and place-making). In essence, community is "a dynamic, multi-dimensional, as well as spatially contextualized environment that both shapes and is shaped by the experiences and practices of older people" (Buffel et al., 2012, p. 27).

Thus, research on the relationship between aging and community has recognized the diverse and critical functions undertaken by community. Rather than being static, community is in fact a vivid organism with versatile functions with which older adults to interact. Time and place interact in the community, providing older adults with a sustainable and interactive ecology (Ahn et al., 2020). This interactive characteristic of aging in community goes hand in hand with a general gerontological understanding that aging is perhaps an evolving adaptation between individuals and their environments (Lawton, 1990; Lawton, Weisman, Sloane, & Calkins, 1997; Moore, 2014).

Greenfield (2012) theorized aging in community from two ecological perspectives: the general ecological model of aging (Lawton & Nahemow, 1973) and bioecological systems theory (Bronfenbrenner & Morris, 2006). Both ecological perspectives focus on dynamic, interactive person–environment processes (Greenfield, 2012). For aging in community, in particular, it is important to recognize the physical and social interactions between individuals and environments, which may facilitate or challenge the overall aging process. In this book, we primarily adopt the social-ecological framework (Bronfenbrenner, 1979) and bioecological systems theory (Bronfenbrenner & Morris, 2006) to understand the dynamics of community for older adults' interactions with HCBS in Shanghai. Specifically, we examine various interactions between older adults and community on different levels, as well as the overall connection among interactions. In addition, we situate the HCBS program and its 13-year-long development in the Jing'an District as manifestations of social changes in urban China.

The ecological framework offers a way to understand the dynamic interrelations among various personal and environmental factors (Bronfenbrenner, 1979), which can be considered a comparatively bounded structure consisting of interacting, interrelated, or interdependent elements that form a whole (Mlinar, 1978). Drawing from natural ecosystems, which are defined as the network of interactions among organisms and between organisms and their environment, social ecology situates a community within an environment, such as health systems, education systems, and economic systems. In particular, the social-ecological framework has been a prominent guide in promoting community health (e.g., Anderson, Scrimshaw, Fullilove, & Fielding, 2003; Cashman et al., 2008; Minkler & Wallerstein, 2008).

The ecological framework pays explicit attention to the social, institutional, and cultural contexts of people–environment relations. This perspective emphasizes the multiple dimensions (e.g., physical environment, social and cultural environment, personal attributes), multiple levels (e.g., individuals, groups, organizations), and complexity of human situations (e.g., cumulative impacts of events over time; Wagemakers, Vaandrager, Koelen, Saan, & Leeuwis, 2010).

The ecological framework also incorporates concepts such as interdependence and homeostasis from systems theory to characterize reciprocal and dynamic person–environment transactions (Stokols, 1996). As such, social ecology goes beyond descriptive analysis, broadening the theoretical scope by stretching the time frame from cross-sectional to longitudinal; extending the levels of social systems, including individual, community, and institutional; and incorporating social–environment interactions and connectedness (Mlinar, 1978). In fact, community-based services provide opportunities for older adults to maintain social connectedness (Chen et al., 2019; Nosraty, Jylhä, Raittila, & Lumme-Sandt, 2015; Scharlach & Lehning, 2013; Wiles, Leibing, Guberman, Reeve, & Allen, 2012). Older adults in Shanghai have also reported that frequent interactions in the HCBS service centers enhanced their life satisfaction (Chen et al., 2019; Ye, Chen, & Kahana, 2017).

1.2 Shanghai: The "Oldest" City in China

Aging trends have posed unprecedented challenges for long-term eldercare in many countries around the world. In particular, China has become the only country on earth with more than 100 million elderly people since 2010 (Population Reference Bureau [PRB], 2010). This number will exceed 200 million in the next 3 years and reach a peak of 300 million by 2025. In 2018, the number of older adults aged 65 and over reached almost 165 million in China, accounting for 11.5% of the total population. By 2030, the number is projected to be almost 250 million, accounting for approximately 20% of the total Chinese population, with more than a 30% economic old-age dependency ratio (United Nations, 2019).

As the first city in China whose aging population (60+)[1] reached 10% in 1979, Shanghai has been the "oldest" city in China ever since (The Central People's Government of the People's Republic of China, 2009). The number of those who were 60 years old and over was 5.03 million, accounting for 34.4% of the total population in 2018 (Shanghai Research Center on Aging, 2019). The life expectancy of Shanghai's registered residents reached 83.63 years (81.25 for men and 86.08 for women) in 2018, the highest in China and higher than all but a few countries in the world (Shanghai Research Center on Aging, 2019). The aging population in Shanghai has increased by 60%, from 3.15 million in 2010. On a special note, beginning in 2012, more than 80% of the increased number of older adults were parents of the only-child generation (Shanghai Research Center on Aging, 2019).

Shanghai provides a good case for examining evolving HCBS for older adults in the Chinese context. With a long history of being the largest and most developed city in China, Shanghai has often been chosen by the Chinese central government as a model city for implementing new policy reforms (Feng et al., 2013). Specifically, since the 1978 Economic Reform in China that transformed the policymaking mechanism from centralization to decentralization (Chu & Chi, 2008), Shanghai's government has implemented a variety of policies to improve older adults' well-being, such as institutional caregiving and community-based eldercare. HCBS began to emerge after 2001 as a result.

In 2001, the Shanghai government and its Civil Affairs Bureau proposed to establish community care centers for older adults in each residential neighborhood[2] (*juwei* in Chinese; Chen & Han, 2016; Shanghai Civil Affairs Bureau, 2001). However, these centers were mainly for recreational purposes. Between 2003 and 2007, the Shanghai government proposed the "90-7-3" eldercare framework highlighted HCBS (Shanghai Government, 2007), which divided older adults into three groups: 90% primarily relying on family caregiving for health care; 7%

[1] In this book, we define the aging population as "adults 60 years old or over." In China, the retirement age is 55 for women and 60 for men. Thus, Chinese statistics and government's welfare policies for the aging population use the age 60 cutoff. We recognize that this is different from the usual age cutoff of 65 years that is widely used by many other countries.

[2] The neighborhood-level administration unit in the Chinese society (Bian, Y., 1997. Bringing strong ties back in: Indirect ties, network bridges, and job searches in China. *American Sociological Review*, *62*, 366–385).

seeking HCBS for care and assistance, in addition to family caregiving; and the remaining 3% possibly needing to move into nursing homes as their healthcare needs exceeded family caregiving capacities (Shanghai Government, 2007). HCBS in Shanghai aims to integrate fragmented geriatric and long-term care resources to support the majority of its older residents (i.e., 97%) who do not require institutional caregiving (Chen & Han, 2016). In 2010, the Shanghai government expanded service coverage and delivery for the increasing number of older adults in the community, mandating each HCBS service center to offer meal service, bathing assistance, home care, and emergency support for those older adults in need (Chen & Han, 2016; Shanghai Civil Affairs Bureau, 2010). In 2011, the State Council of China proposed "Establishing a Social Support System of Elder Care (2011–2015)" to promote community-based eldercare across the country (State Council of China, 2012).

1.3 Our Study: The HCBS Program in the Old Jing'an District

The old Jing'an District[3] was the first to introduce HCBS programs for its aging population because it was the "oldest" district in Shanghai. In 2013, 30.6% of its 2.96 million people were over 60 years old and 53% of these older adults were female. Among these older adults, 20,000 were at least 80 years old, accounting for 22.1% of the total older population in the district (Shanghai Research Center on Aging, 2014).

The old Jing'an District is located in the center of Shanghai, an area of 7.6 km^2 with five sub-districts (*jiedao* in Chinese) and 72 residential communities (*juwei*; see Figs. 1.1 and 1.2). In 2013, approximately 2.96 million people lived in this district, with the highest GDP per capita (US$40,000) in Shanghai (Jing'an Yearbooks Editorial Department, 2014; Shanghai Statistics Bureau, 2014). As one of the most developed areas in Shanghai, the old Jing'an government has spent 999 billion CNY (about US$145 billion) in government subsidies to civil affairs, which has included the HCBS program for older adults since 2001 (Shanghai Statistics Bureau, 2014).

[3] The current Jing'an District in Shanghai consists of the old Jing'an District and the old Zhabei District. The two districts merged in November 2015.

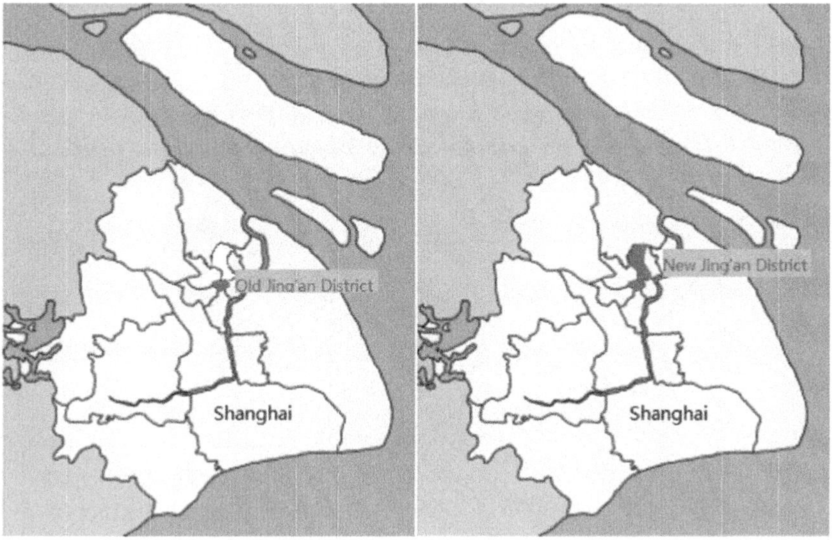

Fig. 1.1 The old Jing'an District and the new Jing'an District

The number of older adults living in the same area exceeded 900,000, accounting for 33% of the total population in the Jing'an District. The density of the elderly population was over 10,000 people/km^2 in 2015 (Jing'an Yearbooks Editorial Department, 2014). According to an official report, more than 13,000 older adults received services from the community centers in 2014 (Jing'an Yearbooks Editorial Department, 2014).

The design of the HCBS program was based on the United Nations Principles for Older Persons (United Nations, 2008), that is, independence, participation, care, dignity, and self-fulfillment. The HCBS program requires each service center to cover 10 types of service:

- Assistance service: Staff helps elderly clients to run errands, including making purchases, picking up packages, or paying for utilities;
- Bathing assistance: Each service center installs shower stalls, and staff members help elderly clients to bathe;
- Companion service: Staff members accompany elderly clients for some outdoor activities;

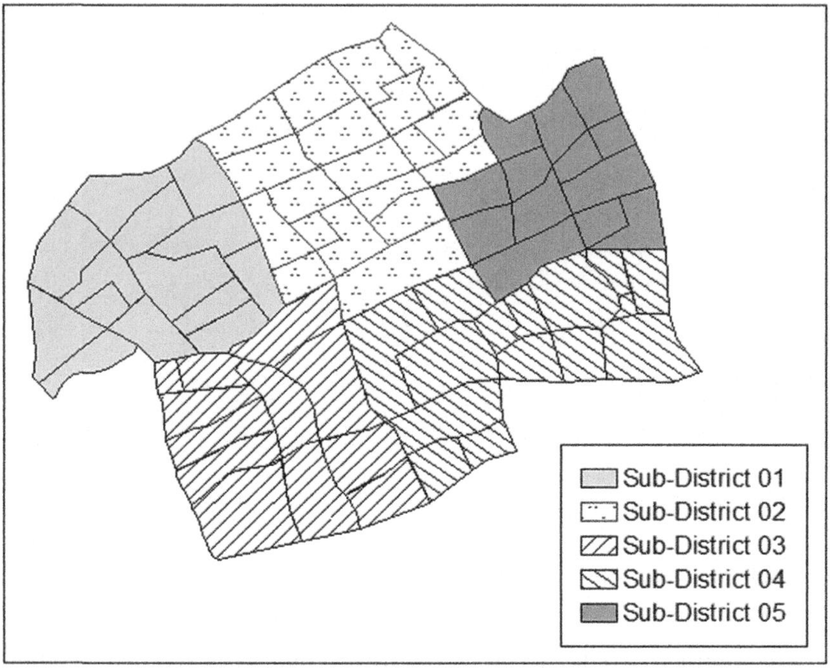

Fig. 1.2 The five sub-districts and 72 residential communities in the old Jing'an District

- Consulting and counseling service: Professional psychotherapists, lawyers, or social workers are invited to service centers to provide psychological counseling and legal consulting for older adults in the neighborhood, regardless if they are elderly clients;
- Meal service: Each service center provides lunch and/or dinner for enrolled elderly clients, with flexible meal plans. Elderly clients can order the number of meals at the beginning of every month or pay per meal as they go. They can also choose to have meals at service center, take-home, or a free delivery to home;
- Emergency service: Staff members help elderly clients to deal with emergencies, such as falls and other medical emergencies;
- Entertainment service: Staff members organize community activities, festivals, arts exhibitions, or sports for elderly clients;

- Education service: Staff members hold lectures (e.g., health promotion) and courses (e.g., calligraphy, computer), or reading clubs for elderly clients;
- Medical service: Staff members invite professional physicians to provide medical consulting for elderly clients;
- Personal care service: Staff members outsource hairdressers and manicurists for elderly clients in the service center.

The HCBS program was designed to serve as many older adults as possible and provide as much proximity as possible. The HCBS program plans to have one service center covers an area that two to three residential neighborhoods are in charge of, and that it takes 1–5 minutes for older adults to walk from their home to the closest center. Following the Shanghai Civil Affairs Bureau, the number of service centers is likely to rise to cover the increasing number of community-dwelling older adults in Shanghai.

All adults aged 60 and over living in the Jing'an District can enroll in the HCBS program. Most program services are free; however, some services, such as meal service and assistance service, charge low fees (e.g., each meal costs about $1.50). Those older adults who have extremely low incomes, live alone, and/or have disabilities receive coupons from the district government to enroll in the program for free.

Between 2011 and 2018, we conducted observations, in-depth interviews, and surveys with elderly clients of the HCBS program to illustrate the dynamic interactions between older adults and their community. We interviewed frontline staff members of the HCBS program to elicit their work experiences and their perspectives on HCBS. We also interviewed the director of the HCBS program and the Aging Department of Jing'an District to trace back the history of the HCBS program and healthcare policy development in Shanghai.

1.4 Overview of the Book

This book consists of seven chapters. The introduction outlines the broader social-ecological development context of HCBS in urban China. It briefly introduces the study design, the book's major goals, and its organization. Chapter 2 presents the social-ecological framework—focusing on interactions among macro, meso, and micro levels—to guide the following analysis of the multilayered interactions between older adults and the HCBS program as well as the community. Chapter 3 illustrates core concepts and contexts of one HCBS program in Shanghai, reviewing its development, including its initial stages, policy implementations, current progress, and evaluation procedures, which have shaped the program services as well as the quality of care. In Chapter 4, we use the geographic information system (GIS) to investigate how the district government has strategically allocated resources to maximize the capacity of each service center of the HCBS program. On the macro level of the community ecology, we analyze the choices of service centers in different geographic locations in relation to the communities they serve. Chapter 5 describes work experiences from frontline staff members' perspectives in the service centers of the HCBS program. This chapter highlights the interactions on both meso and micro levels of the community ecology for older adults as well as staff members in the HCBS program. Chapter 6 provides a detailed analysis of the micro-level interactions in the community ecology to show how meal service can help to improve elderly clients' life satisfaction, both instrumentally and socially. Chapter 7 concludes and reflects on our analysis of the community ecology and various interactions within the ecology in the prior chapters. We propose future policy and practice directions to improve HCBS for older adults in urban China.

References

Ahn, M., Kwon, H., & Kang, J. (2020). Supporting aging-in-place well: Findings from a cluster analysis of the reasons for aging-in-place and perceptions of well-being. *Journal of Applied Gerontology, 39*, 3–15.

Anderson, L. M., Scrimshaw, S. C., Fullilove, M. T., & Fielding, J. E. (2003). The community guide's model for linking the social environment to health. *American Journal of Preventive Health, 24*(S3), 12–20.

Blanchard, J. (2013). Aging in community: Communitarian alternative to aging in place, alone. *Generations, 4*, 6–13.

Bronfenbrenner, U. (1979). *The ecology of human development: Experiments by nature and design*. Cambridge, MA: Harvard University Press.

Bronfenbrenner, U., & Morris, P. A. (2006). The bioecological model of human development. In W. Damon & R. M. Lerner (Eds.), *Handbook of child psychology, Vol. 1: Theoretical models of human development* (6th ed., pp. 793–828). New York: Wiley.

Buffel, T., Verté, D., De Donder, L., De Witte, N., Dury, S., Vanwing, T., & Bolsenbroek, A. (2012). Theorising the relationship between older people and their immediate social living environment. *International Journal of Lifelong Education, 31*(1), 13–32. https://doi.org/10.1080/02601370.2012.636577.

Callahan, J. J. (Ed.). (1993). *Aging in place*. Amityville, NY: Baywood.

Cashman, S. B., Adeky, S., Allen, A. J., Corburn, J., Israel, B. A., Montaño, J., … E. Eng. (2008). The power and the promise: Working with communities to analyze data, interpret findings, and get to outcomes. *American Journal of Public Health, 98*(8), 1407–1417.

Center for Disease Control and Prevention. (2013). *Healthy places terminology: Aging in place*. Retrieved from http://www.cdc.gov/healthyplaces/terminology.htm.

Chen, L., & Han, W. J. (2016). Shanghai: Front-runner in community-based elder care in China. *Journal of Aging & Social Policy, 28*(4), 292–307. https://doi.org/10.1080/08959420.2016.1151310.

Chen, L., Ye, M., & Kahana, E. (2019). A self-reliant umbrella: Defining successful aging among the old-old (80+) in Shanghai. *Journal of Applied Gerontology*. https://doi.org/10.1177/0733464819842500.

Chipuer, H. M., & Pretty, G. M. H. (1999). A review of the Sense of Community Index: Current uses, factor structure, reliability, and further development. *Journal of Community Psychology, 27*(6), 643–658.

Chu, L., & Chi, I. (2008). Nursing homes in China. *Journal of American Medical Director Association, 9*, 237–243.

Davey, J., Nana, G., de Joux, V., & Arcus, M. (2004). *Accommodation options for older people in Aotearoa/New Zealand*. Wellington, New Zealand: NZ Institute for Research on Ageing/Business & Economic Research Ltd, for Centre for Housing Research Aotearoa/New Zealand.

Feng, Q., Zhen, Z., Gu, D., Wu, B., Duncan, P. W., & Purser, J. L. (2013). Trends in ADL and IADL disability in community-dwelling older adults in Shanghai, China, 1998–2008. *Journals of Gerontology. Series B, Psychological Sciences and Social Sciences, 68*(3), 476–485.

Greenfield, E. A. (2012). Using ecological frameworks to advance a field of research, practice, and policy on aging-in-place initiatives. *The Gerontologist, 52*(1), 1–12.

Greenfield, E. A., Black, K., Buffel, T., & Yeh, J. (2019). Community gerontology: A framework for research, policy, and practice on communities and aging. *The Gerontologist, 59*, 803–810. https://doi.org/10.1093/geront/gny089.

Jing'an Yearbooks Editorial Department. (2014). *Jing'an Almanac 2014*. Retrieved from http://www.jingan.gov.cn/jagl/janj/2014janj/janj2014.html [Chinese].

Lawler, K. (2001). *Aging in place: Coordinating housing and health care provision for America's growing elderly population*. Washington, DC: Joint Center for Housing Studies of Harvard University & Neighborhood Reinvestment Corporation.

Lawton, M. P. (1990). Knowledge resources and gaps in housing for the aged. In D. Tilson (Ed.), *Aging in place* (pp. 287–309). Glenview, IL: Scott Foresman.

Lawton, M. P., & Nahemow, L. (1973). Ecology and the aging process. In C. Eisdorfer & M. P. Lawton (Eds.), *Psychology of adult development and aging* (pp. 619–674). Washington, DC: American Psychological Association.

Lawton, M. P., Weisman, G. D., Sloane, P., & Calkins, M. (1997). Assessing environments for older people with chronic illness. *Journal of Mental Health and Aging, 3,* 83–100.

Minkler, M., & Wallerstein, N. (2008). *Community-based participatory research for health: From processes to outcomes* (2nd ed.). San Francisco, CA: Jossey-Bass.

Mlinar, Z. (1978). A theoretical transformation of social ecology: From equilibrium to development. In Z. Mlinar & H. Teune (Eds.), *The social ecology of change: From equilibrium to development*. Thousand Oaks, CA: Sage.

Moore, K. D. (2014). An ecological framework of place: Situating environmental gerontology within a life course perspective. *International Journal of Aging, 79,* 183–209.

Nosraty, L., Jylhä, M., Raittila, T., & Lumme-Sandt, K. (2015). Perceptions by the oldest old of successful aging, vitality 90+ study. *Journal of Aging Studies, 32,* 50–58.

Palley, H. A. (Ed.). (2009). *Community-based programs and policies: Contributions to social policy development in health care and health care-related services*. New York, NY: Routledge.

Population Reference Bureau (PRB). (2010, July). *Today's research on aging: China's rapidly aging population. Program and policy implications* (Issue 20). Washington, DC: Population Reference Bureau.

Putnam, R. D. (2000). *Bowling alone: The collapse and revival of American community*. New York, NY: Simon & Schuster.

Scharlach, A. E., & Lehning, A. J. (2013). Ageing-friendly communities and social inclusion in the United States of America. *Ageing & Society, 33*(1), 110–136.

Scharlach, A. E., & Moore, K. D. (2016). Aging in place (Chapter 21). In V. L. Bengtson & R. A. Settersten (Eds.), *Handbook of theories of aging* (pp. 407–425). New York, NY: Springer.

Shanghai Civil Affairs Bureau. (2001). *To adapt to the increasing aging population in Shanghai, home- and community-based services for older adults should be promoted to the entire city*. Retrieved from http://www.shweilao.cn/cms/cmsDetail?uuid=fcd27c92-ca9f-4908-b9a0-ac099f8e1c54 [Chinese].

Shanghai Civil Affairs Bureau. (2010). *About Shanghai localized "regulations of elderly home care in community" explanations*. Retrieved from http://mzj.sh.gov.cn/gb/shmzj/node8/node890/userobject1ai27337.html [Chinese].

Shanghai Government. (2007). *Shanghai government dedicates to "90-7-3" elder-care framework*. Retrieved from http://www.shanghai.gov.cn/shanghai/node2314/node2315/node4411/userobject21ai238364.html. [Chinese].

Shanghai Research Center on Aging. (2014). *Shanghai aging population statistics 2013*. Retrieved from http://www.shshjs.gov.cn/shjs/node5/node34/u1a33197.html [Chinese].

Shanghai Research Center on Aging. (2019). *Shanghai aging population statistics 2018*. Retrieved from http://mzj.sh.gov.cn/gb/shmzj/node4/node10/n2708/u1ai47303.html [Chinese].

Shanghai Statistics Bureau. (2014). *Shanghai Statistical Yearbook 2014*. Retrieved from http://tjj.sh.gov.cn/html/sjfb/201701/1000200.html [Chinese].

State Council of China. (2012). *Plan to establish a social support system of eldercare (2011–2015)*. Retrieved from http://www.shmzj.gov.cn/gb/shmzj/node8/node15/node55/node230/node246/u1ai31182.html [Chinese].

Stokols, D. (1996). Translating social-ecological theory into guidelines for community health promotion. *American Journal of Health Promotion, 10*(4), 282–298.

The Central People's Government of the People's Republic of China. (2009). *The life expectancy in Shanghai reached 81.28 years old*. Retrieved from http://www.gov.cn/jrzg/2009-04/20/content_1290887.htm [Chinese].

Thomas, W., & Blanchard, J. (2009). Moving beyond place: Aging in community. *Generations, 33*(2), 12–17.

United Nations. (2008). *United Nations principles for older persons*. Geneva: Switzerland.

United Nations. (2019). *World population ageing: Highlights*. Geneva: Switzerland.

Vasunilashorn, S., Steinman, B. A., Liebig, P. S., & Pynoos, J. (2012). Aging in place: Evolution of a research topic whose time has come. *Journal of Aging Research, 2012*. Article 120952.

Wagemakers, A., Vaandrager, L., Koelen, M. A., Saan, H., & Leeuwis, C. (2010). Community health promotion: A framework to facilitate and evaluate supportive social environments for health. *Evaluation and Program Planning, 33*(4), 428–435.

Wiles, J. (2005). Conceptualizing place in the care of older people: The contributions of geographical gerontology. *Journal of Clinical Nursing, 14*, 100–108.

Wiles, J. L., Leibing, A., Guberman, N., Reeve, J., & Allen, R. E. (2012). The meaning of "aging in place" to older people. *The Gerontologist, 52*(3), 357–366.

World Health Organization. (2007). *Global age-friendly cities: A guide*. Geneva: Switzerland.

Wu, B., Carter, M. W., Goins, R. T., & Cheng, C. (2005). Emerging services for community-based long-term care in urban China: A systematic analysis of Shanghai's community-based agencies. *Journal of Aging & Social Policy, 17*(4), 37–60.

Ye, M., Chen, L., & Kahana, E. (2017). Mealtime interactions and life satisfaction among older adults in Shanghai. *Journal of Aging and Health, 29*(4), 620–639. https://doi.org/10.1177/0898264316641080.

CHAPTER 2

Community as a Care Ecology

This chapter introduces the theoretical framework of this book. First, it reviews the history of social-ecological theory. Then it focuses on Urie Bronfenbrenner's ecological theory (Bronfenbrenner, 1977) and Emily Greenfield's ecological theory (Greenfield, 2012) to discuss the macro, meso, and microsystems in the community care ecology and how to understand HCBS in China from the social-ecological perspective.

2.1 A Brief History of Social-Ecological Theory

The original social-ecological theory was introduced by Bronfenbrenner in the late 1970s (Bronfenbrenner, 1977). This theory conceptualizes human development as "the phenomenon of continuity and change in the bio-psychological characteristics of human beings" (Bronfenbrenner & Morris, 2007, p. 795). Contemporary psychologists such as Bandura (2009) also believed that the environment affected the development of individuals (Zimmerman & Schunk, 2003). However, many scholars did not provide a clear description of the environment in which individuals develop. Bronfenbrenner's social-ecological theory, on the contrary, emphasizes that the developing individual is nested in a series of environmental systems that affect each other. This theoretical contribution captures a detailed conceptual description of the environment in which individuals develop. Although Bronfenbrenner's (1977) original social-ecological theory mainly described children's development, he argued

L. Chen and M. Ye, *Community Eldercare Ecology in China*,
https://doi.org/10.1007/978-981-15-4960-1_2

that human development studies should focus on the environmental setting and the process of adjustment "throughout the lifespan, between the growing human organism and the changing environments in which it actually lives and grows" (Bronfenbrenner, 1977, p. 513).

Since then, Bronfenbrenner's social-ecological theory has changed the way psychologists appraise child development (Tudge et al., 2016). In fact, Bronfenbrenner himself argued that his work was "an evolving theoretical system for the scientific study of human development over time" (Bronfenbrenner & Morris, 2007, p. 793). Before Bronfenbrenner elaborated his theory during the mid-twentieth century, most child development scientists emphasized particular environmental impacts, such as varying institutional (e.g., family) or social and structural (e.g., socioeconomic class) environments, on the differences of children's physical, psychological, and social development (Waters & Crandall, 1964).

Following Bronfenbrenner, child development scientists started to observe the comprehensive environmental system that contains many different levels and types of environmental factors that may affect a child's development (Tudge et al., 2016). For instance, many studies have examined how different levels of the environment in the school system influence a child's development (Langille & Rodgers, 2010; Turner, 2019). Espelage and Swearer (2010) employed a social-ecological framework to analyze how micro-level interactions (i.e., peer and parents' relationship) influenced bullying among teenagers.

Scholars have further adopted an interdisciplinary approach to integrate social-ecological theory with other theories to advance the understanding of multiple dimensions of individuals' environments (Hayes, O'Toole, & Halpenny, 2017; Shelton, 2018). For example, Hayes et al. (2017) pointed out that cultures, values, and beliefs have significant and direct influences on pedagogical approaches in children's education and development. Since the middle of the twentieth century, a child-focused approach has become popular, and teachers tend to pay attention to internal motivations and children's own characteristics instead of using external rewards to teach students (Hayes et al., 2017; Kohn, 1999; Schunk, 2012). This reflects a more sociological approach to education: cultures and beliefs influence teachers' instruction and expectations for students which, in turn, affect child development. The relationship between teachers and students can be shifted by reforming pedagogical philosophy instead of relying on traditional approaches to classroom management (i.e., reward vs. punishment) that seek individual behavioral change.

The content of social-ecological theory has been applied to many aspects of older adults' life, physical functioning, and health-seeking behaviors. For example, Ostrom (2014) employed social-ecological theory to analyze collective actions, observing how institutional regulations and arrangements enable people to work together and solve problems. Greenfield (2012) combined social-ecological theory and Powell's (1990) residential environment model to conceptualize programs of aging in place and analyze the advantages and disadvantages of these programs. Glanz, Rimer, and Visawanath (2015) used social-ecological theory to study personal health behaviors. Their study explored how culture and social media affect health communication and behaviors (Glanz et al., 2015). Moran and colleagues (2016) studied how to understand changes in health behaviors. From individual-level factors (e.g., families and peers), exosystem-level context (e.g., neighborhood and institutions), and macro-level content (e.g., cultural norms and values), Moran and colleagues (2016) developed ecological measures and methods to capture comprehensive health behaviors beyond the simple individual level. Ma, Chan, and Loke (2017) used social-ecological theory to understand barriers to sex workers' health services. In their study, those barriers implicate multiple sociological factors, including social support, service information, social stigma, and healthcare policy (Ma et al., 2017). Bowlby and McKie (2019) developed the concept of "care ecology" (p. 532) based on social-ecological theory and observed individuals' interactions to understand formal and informal care support, services, and challenges in this ecology. Their study recommended building new forms of community support to cover the scope of formal and informal care in the care ecology. Galafassi and colleagues (2017) used social-ecological theory in an intervention study to provide workshops for representatives of communities, government, and NGOs in Kenya to discuss local issues, and found that the workshops indirectly influenced management and policymaking.

Thus, the purpose of social-ecological theory is to depict the relationship between people's behaviors and their physical and social situations. Although many studies have only partially tested Bronfenbrenner's theory without doing a comprehensive analysis—using all three levels of systems as well as a time-frame system (Tudge et al., 2016)—it is still a pivotal theoretical framework for understanding individuals' behaviors, relationships, and development; in its 40 years of evolution, social-ecological theory dominates many fields and topics, including child development (e.g., Espelage & Swearer, 2010), collective actions (e.g., Ostrom, 2014), health behaviors (e.g., Glanz et al., 2015), public health services (e.g., Ma et al., 2017), caregiving and support (e.g., Bowlby & McKie, 2019), social institutions (e.g., Greenfield, 2012), and political science (e.g., Görg et al., 2017).

2.2 Bronfenbrenner's Social-Ecological Theory

Unlike other early psychologists in the mid-twentieth century who conducted research in laboratory settings to identify and describe children's behaviors (Reifsnider, Gallagher, & Forgione, 2005), Bronfenbrenner observed children in a real-world setting and treated the environment as a complex system that involves specific social, cultural, and institutional practices (Shelton, 2018). These external influences and practices work together to provide the settings and experiences that contribute to human development. According to Bronfenbrenner, the natural environment is the main determinant of human development—a fact that is often overlooked by researchers in artificially designed laboratories (Bronfenbrenner, 1979).

Bronfenbrenner's work presents a complex and holistic system, entailing different layers of environment and the relationships among these layers, whose structure can be compared to that of an onion or a Russian doll (Parrott & Meyer, 2012). The structure of "each inside the next" contributes to "a set of nested structures," characterizing every environment from small, such as a family or a school, to large, such as cultural customs, policies, or an entire economy (Bronfenbrenner, 1977, p. 3). The developing individual is in the middle of, or nested within, several external environmental systems, ranging from the direct, micro-level environment (e.g., family) to the indirect, macro-level environment (e.g., policy). Each system interacts with other systems and individuals, affecting many important aspects of the development of individuals. Bronfenbrenner differentiated among these systems and named them microsystem, mesosystem, exosystem, and macrosystem. Later, he added a time-frame system—the chronosystem—to this theory to measure the influence of time and historical context on a child's development

(Bronfenbrenner, 2001). The following section will introduce each of these five systems in detail.

2.2.1 *Microsystem*

The microsystem represents the immediate environment in which a person lives (Bronfenbrenner, 1977; Schlechter, 2015, 2018). It is a constantly changing and developing system and is the innermost layer of the environmental system. It is a space in which the person has direct interactions with at least one other actor, typically family members and close friends. The person has a direct connection to everything in this environment; therefore, almost everything in such a close and personal setting can have an impact on the person. The physical environment can be a room or a house.

The scope of the micro-level system can change according to the development of a person. For example, for most babies, the microsystem is limited to the family. As infants continue to grow, their scope of activities continues to expand. Kindergarten, school, and peer relationships continue to be incorporated into the infant's microsystem. For students, the school is the microsystem that has a significant impact on them outside the home (Espelage & Swearer, 2010). For adults, a microsystem can be their workplace or small interest groups (Ettner & Grzywacz, 2001). For older adults, a microsystem can be a senior center that they visit every day (Hickerson et al., 2008).

In a micro-level system, the relationships between actors are important. A relationship in a micro-level system has two features: direct interactions and two-way interactions. Direct interactions mean that in microsystem interactions are direct and frequent. The environment of direct interactions, such as the characteristics of these close relationships, also has a direct influence on the person. Bronfenbrenner emphasizes that in a micro-level system, the interaction between people is a mutual one (Shelton, 2018). That is, how the person reacts to other people in the microsystem reciprocally affects how the others treat that person (Shelton, 2018).

For example, a child's microsystem includes their parents (Bronfenbrenner, 1977; Langille & Rodgers, 2010). The baby's cry attracts its mother's attention and affects her breastfeeding behavior. If the mother can feed the baby in time, the baby will stop crying. When interactions between children and adults are well established and occur often, they can have lasting effects on the child's development. The more positive the stimulation, the better the foundation the child's brain will have (Bronfenbrenner, 1979). In other words, not only do adults influence children's responses, but children also influence adults' responses. Children's personalities and abilities also influence adult behavior (Shelton, 2018; Tudge et al., 2016).

The mutual relationship between parents and the infant is very important for a child's development. Previous studies have demonstrated that how parents interact with their children can affect a child's organ development. To put it metaphorically, these parent–child interactions are like water in which the child is swimming (Dannefer, 2008). The development of a human organ is flexible and sensitive to the social environment (Dannefer, 2008). A person's genetic background can provide a blueprint for the organ, but his or her own experiences are important in shaping the process of the organ's development (Halldorsdottir & Binder, 2017). The developing brain needs billions of brain cells to send electrical signals to communicate with each other (Stiles, Brown, Haist, & Jernigan, 2015). Electrical signals are usually stimulated by a child's interaction with its environment, such as parents' touching and talking. These interactions set a base for the child's development. Repeated simulation can make the links more efficient and faster (Stiles et al., 2015). Stiles et al. (2015) demonstrated that a child who experiences neglect in early childhood has a smaller brain than one who does not face neglect but receives support and caring attention. Because any later connections are built by the first forms, as in the construction of a house, the basic foundations created by the microsystem are essential to a child's future learning behaviors and health status (Posner & Rothbart, 2018).

Microsystems grow more complex throughout the life course. As a care receiver, an older adult's microsystem can be his or her home and the family members with whom he or she interacts daily. Usually, spouses are the main caregivers for older adults at home (Ye, DeMaris, & Longmore, 2018). Since the relationship between caregivers and care receivers is a two-way interaction, the caregivers' status is also important for older adults (Kahana & Young, 1990). For example, spouses' health status is important for older adults' quality of life. Lee and DeMaris (2007) demonstrated that older adults tend to report an increase in depression when their spouses' health was poor.

The neighborhood is another micro-level system for older adults when they interact with their neighbors daily. In a previous study (Ye & Chen,

2014), we showed that by participating in community volunteer activities and enjoying older adult services provided by the community, older adults developed a sense of belonging to their community and feelings of pride, which enhanced their well-being. Hickerson et al. (2008) employed a social-ecological model to observe how American senior centers in communities promoted physical activities among older adults through peer interactions. Their study pointed out that although senior centers may have rich physical-activity resources for older adults, they do not necessarily participate in these activities if they lack peer recognition and support.

Older adults may also be caregivers for their offspring. For example, if they are taking care of or living with a grandchild, then the grandchild also has an impact on the grandparents' life, whether it is because the grandchild can help to take care of the grandparents or because the grandchild needs support from the grandparents (Kahana et al., 2015). In the Chinese context, grandparents are often expected to look after their grandchildren, and their interactions with their grandchildren also influence their physical and psychological well-being (Xu, 2019).

2.2.2 Mesosystem

The mesosystem represents the interactions among different microsystems, such as family members and friends (Ferguson & Evans, 2019). Various microsystems may work independently or interdependently. One system may influence another, eventually having an indirect impact on the person (Greenfield, 2012).

Bronfenbrenner argued that childhood development can be optimized if there is a strong positive connection among microsystems. In contrast, nonpositive connections between microsystems can have negative consequences (Bronfenbrenner, 2001). In other words, a well-functioning mesosystem is crucial to the success of its component microsystems. For example, the child's social world and physical environment extend beyond its parents and the home. The child attends school and other institutions and establishes relationships there. Teachers also see the child every day. A child's performance and development outside of the family depends not only on people associated with the institutions but also on the interrelationships between the parents and the institutions. The connection among the parents, the institutions, and the people working in the institutions can be seen as a mesosystem of the child (Espelage & Swearer, 2010).

If the parents do not develop their relationship with the child within the family, the child does not learn how to build successful and intimate and relationships. And in school, it will not be easy for the child to establish close friendships with classmates; it will also affect the relationship between the teacher and the child (Crosnoe, 2004; Espelage & Swearer, 2010). Whether the parents are involved in institutional activities and interactions with staff, whether the teachers and the parents collaborate to educate the child, whether the parents like the child's peer groups—these considerations are all central components of the child's development (Langille & Rodgers, 2010). If parents can play positive roles in their children's various micro-level systems—for example, by establishing positive relationships with schools, teachers, and classmates—their children will have optimal social conditions for development (Crosnoe, 2004; Espelage & Swearer, 2010).

Espelage and Swearer (2010) used the social-ecological model to design interventions in schools to prevent bullying among young children. Their study reveals that parents can dramatically influence young children's behaviors and experiences of violence at school, while the educational environment has a significant impact on shaping children's development. Langille and Rodgers (2010) used the social-ecological model to explore the influence of physical activities in school and found that the physical and social environments have a strong impact on physical-activity strategies.

For older adults, the mesosystem—multiple microsystems and the relationships among them—are important for their quality of life. For example, scholars have found that the relationship between adult children and staff members in a nursing home is important to the quality of life of their elderly parents who live there (A. R. Roberts & Ishler, 2018; T. Roberts & Bowers, 2015). For staff members in nursing homes, it is beneficial to have a positive relationship with the residents' adult children, one that entails both fruitful support from and regular communication with the adult children. This positive relationship can inform the staff members' understanding of and sympathy for the older adults in the nursing home, and thus improve the services and treatment that they receive from the staff members, resulting in a better quality of life (A. R. Roberts & Ishler, 2018; T. Roberts & Bowers, 2015).

Garza (2018) explored the barriers to utilization of HCBS among older adults and pointed out that mesosystem-level factors are often ignored by researchers. Indeed, the mesosystem plays a very important role in increasing or decreasing HCBS usage because the delivery of HCBS services is a complex system involving multiple levels. She found that an older adult may be involved with several networks, such as a Medicare/Medicaid program, National Aging Network organizations, long-term services and support (LTSS) programs, and HCBS programs. The relationships among these networks can heavily influence HCBS accessibility for older adults. When they interact with these systems, their family members are also involved with the experience of these services, and the relationship between older adults and their family members can affect whether they use or have access to HCBS programs. Because all levels of relationships are mingled together, "changes in one system would influence interactions within the other systems… as such, singular factor interventions implemented in one level without consideration for possible outcomes in other levels may lead to unintentional outcomes, such as barriers to HCBS" (Garza, 2018, p. 99).

2.2.3 Exosystem

An exosystem is a large system that contains at least one mesosystem (Bronfenbrenner, 1977). The difference between mesosystem and exosystem is that the mesosystem focuses on relationships between micro-level systems while the exosystem focuses on a larger system that contains micro-level systems (Garza, 2018). The exosystem usually refers to a setting that does not directly focus on the person as an active agent (Holden, 2019). The person often does not operate in the exosystem directly. Nor do the person and the exosystem interact frequently or even daily. There is usually not an intimate relationship between the two systems, but the exosystem can have an impact on the microsystem and indirectly influence the person (Holden, 2019). For example, whether a local educational system has a disability office can have an impact on the child's development. The disability office does not need to interact directly with the child daily, but its support can have an impact in the classroom (Kahana et al., 2015; Vaccaro & Kimball, 2019). It can provide training for teachers and parents to care for the child; it can also provide facilities that offer specific equipment for children who are disabled so they are able to attend a regular school and achieve academic success with the assistance of this support (Kahana et al., 2015). Another example is a school that offers parenting classes (Vaccaro & Kimball, 2019). The children are not active participants in the classes, but the classes will have an indirect impact on their development through the influence of the parents. Sometimes external networks, such as the work environment, can also encompass and shape the microsystem (Vaccaro & Kimball, 2019). For example, discrimination against women in the workplace can create a toxic environment that exacerbates the stress of the mother and thus has a harmful impact on the microsystem of the child (Casper, Vaziri, Wayne, DeHauw, & Greenhaus, 2018; Zhang, Cubbin, & Ci, 2019).

Likewise, for older adults, if the main caregivers are their adult children, then the work environment of their adult children can indirectly affect the quality of their own lives. For example, if the adult child does not get a promotion or needs to commute to another city for work, these professional frustrations and strains can exert a detrimental influence on the older adults' lives (He et al., 2016). Another more structural or exosystem example is the global financial crisis, which affected the global economy and the work and income of many adult children; older adults who rely on their adult children have experienced corresponding changes. For example, because of increased work pressure, adult children are unable to provide the same level of care to their aging parents (Zeng, Chen, Wang, & Land, 2015). Chan (2006) pointed out that after the 1998 Asian financial crisis, people in Hong Kong faced increased job uncertainty and decreased self-reliance. The crisis posed significant "threats to family integrity," as only 30% of the working population could support their elderly parents; therefore, older adults had to rely more on public welfare instead of family care (Chan, 2006, p. 219).

Another type of exosystem involves organizations that do not help older adults directly but provide services for institutions or staff members who take care of older adults. For example, HCBS programs can be seen as an exosystem for community-dwelling older adults, as the health of older adults is indirectly affected by these community programs (Garza, 2018; Greenfield, 2012). Garza (2018) found that poor communication between HCBS programs and staff members may result in barriers to accessing services among older adults. Also, local training programs can have a significant influence on the number and the quality of staff members in long-term care programs and thus benefit older adults' lives (Surr, Smith, Crossland, & Robins, 2016).

2.2.4 *Macrosystem*

The macrosystem, the largest environment a person has, contains the three previously described systems: the microsystem, mesosystem, and exosystem (Bronfenbrenner, 1977). It encompasses both the immediate people and places and the distant people and places that can affect an individual. It includes people, places, and organizations that the individual interacts with but not often or intimately (Bronfenbrenner, 1977). A person lives in a macrosystem and all the other systems are affected by it. The macrosystem entails a broad network of social arrangements and settings such as the economy, culture, values, and political systems. For example, Townsend and colleagues (2016) coded political and religious news of Northern Irish newspapers to measure the macrosystem. Individuals usually do not interact with the macrosystem directly because it operates at a distal level, but it can influence the exosystem or microsystem directly. Distal level means that the system functions beyond individuals' control and does not involve individuals on a daily basis. Policy and culture are usually distal-level systems. In other words, the macrosystem contributes indirectly to the influences on individuals through other systems (Krieger, 2008).

In China, the latest law on Protection of the Rights and Interests of Older Adults mandates that adult children assume responsibility for supporting their elderly parents (The Standing Committee of the National People's Congress, 2019). In 2017, the updated law also specified that adult children not only have responsibility for financial support but also need to provide emotional support for their elderly parents (The Standing Committee of the National People's Congress, 2019). Because of the authority of Confucian values, filial piety is also a powerful social and psychological force that shapes the relationship between elderly parents and their adult children (X. Chen, 2015; Dong, Zhang, & Simon, 2014; Guo, Xu, Liu, Mao, & Chi, 2016; Simon, Chen, Chang, & Dong, 2014; W. Sun et al., 2015; Wang, Chen, & Han, 2014) These cultural norms prescribe that adult children take care of their elderly parents' health, obey their parents, and fulfill their parents' needs. However, fulfilling their parents' needs goes beyond just taking care of the parents' physical and mental health. For example, some adult children may purchase a house close to their parents because their parents want to live close to their children (Ye, Chen, & Peng, 2017).

The eldercare system in China has also changed because of social changes or, in Bronfenbrenner's terms, the macrosystem. In particular, health care has changed since the Economic Reform (T. Liu & Sun, 2016). Originally, the Government Council of the People's Republic of China promulgated the Labor Insurance Regulations in 1951 (revised in 1953 and 1956), which basically achieved full insurance coverage, including elderly pensions in all cities and towns (T. Liu & Sun, 2016). In 1986, the implementation of the labor contract system established the endowment insurance system for labor contract workers, establishing a social endowment insurance system jointly funded by the state, companies, and individuals (T. Liu & Sun, 2016). By the end of 2016, the average national monthly pension for retirees in cities was 2400 yuan (about 350 US dollars) per person (Xinhua News Agency, 2019). In 2019, the Chinese government has to transfer $21 billion worth of shares in state-owned enterprises into the "National Social Security Fund" to raise the retirees' pensions (*China Daily*, 2019). Chinese older adults spend 15% of their annual income on travel, which has a great impact on the global tourist market (Choi & Fong, 2017). Thus, the spending power of older adults has continuously improved (Choi & Fong, 2017).

Since the 40-year Economic Reform in China, the government has been able to provide more services for older adults than before. By the end of 2017, eldercare services had been expanded to 155,000 agencies offering various types of eldercare services, including eldercare institutions and community care service facilities (Xinhua News Agency, 2018). Chinese universities and senior centers have created many programs and activities to attract older adults. For example, according to China Daily, in addition to offering courses such as calligraphy, art, dance, and Tai Chi, senior universities (i.e., universities for seniors) and service centers also have reading rooms and entertainment rooms to enrich the daily lives of older adults (*China Daily*, 2018).

2.2.5 *Chronosystem*

The chronosystem reflects the influence of history and time on a person's development (Bronfenbrenner, 1977). Social-ecological theory conceptualizes time as a frame of reference for studying psychological changes during individual growth (Bronfenbrenner, 2001). It contains life-span perspectives and investigates how the timing of an event (e.g., college, marriage, retirement) affects a person's development (Carstensen, 1995).

For example, over time, the microsystem environment of children's survival constantly changes, and children will adjust their behaviors according to the external environment. Bronfenbrenner called this kind of environmental change "ecological change," and each change is a stage of individual life development (Shelton, 2018). This idea is similar to life-span theory (Carstensen, 1992, 1995; Carstensen, Isaacowitz, & Charles, 1999), whose basic analytical paradigm is to understand the life course of an individual as a sequence of multiple life events, such as going to school, adolescence, participation in work, marriage, and retirement. If the same set of life events sequence differently, the impact on a person's life can be completely different (Carstensen, 1995).

Time also represents how historical events and periods have an impact on a person's development (Bronfenbrenner, 1979; Elder, 1994; Shelton, 2018). A historical event can provide a setting for a person's development and pervade all the environmental systems that surround the person (Elder, 1994). The chronosystem involves patterns of environmental events and life transitions (George, 1993). It focuses mainly on how the macrosystem interacts with time and how that interaction influences transitions in people's lives (George, 1993). For example, a child of the Great Depression has experienced different developmental process compared with a child who has never lived during an extended period of scarcity and deprivation. A child who experienced the Great Depression might suffer from lack of nutrition, social resources, and job opportunities (Clausen, 1993; Elder, 1994).

For older adults, time also influences how they manage their later lives. For example, traditionally, most of the older adults in China relied on their families to take care of them (Fei, 1939). Because women usually did not work outside of the family, they assumed the largest responsibility for tending to other family members, especially females, at home (Fei, 1939). Nowadays, however, because of the increase in the number of dual-labor families (i.e., both husband and wife have jobs outside the home) and in-migration (e.g., move to another city to work), many older adults cannot depend on their adult children for daily care (Ye & Chen, 2014).

The one-child policy has also changed the structure of the population and the caregiving system in China (Barrows, 2016). Studies have shown that the range of fulfilling parents' needs (i.e., filial piety) is broader than providing instrumental care for parents (L. Chen, Ye, & Kahana, 2019; Sun, 2017; Ye & Chen, 2014). Y. Sun (2017) observed that in modern urban China, especially among those born under the one-child policy, people thought that building intimate intergenerational relationships instead of obeying parents' authority was good for achieving filial piety. Y. Sun's (2017) study argued that filial piety defines a moral virtue of a hierarchical parent–child relationship. In the future, because of the decrease in birth rates and marriage rates, and the development of science and technology, many older adults may have to rely on themselves, neighborhoods, communities, society, and new technologies to support them in their elderly years (L. Chen et al., 2019; Y. Sun, 2017; Ye & Chen, 2014).

In China, the social eldercare system has shifted from relying only on family care to a system that has home-based, community-based, institutional supplemental, and medical support programs. In the past, both elderly parents and their adult children felt ashamed of sending elders to a nursing home. But now, because of the work pressure of the adult children and the innovative concept of eldercare (Ye & Chen, 2014), the number of older adults who choose to go to a nursing home has increased (L. Chen & Ye, 2013).

Bronfenbrenner's social-ecological theory offers insights into how external environments influence human psychological development (Bowlby & McKie, 2019; Galafassi et al., 2017; Ma et al., 2017; Turner, 2019). It provides a framework to analyze multiple layers of context and how these layers influence a person's development. It can also be combined with other theories and models to explain human development (Greenfield, 2012). Methods from psychological, social, and organizational fields can coalesce and find their positions in the social-ecological approach. Social-ecological theory can integrate other theories and thus offers a complex model to analyze the environmental and policy factors at different levels, thus yielding a comprehensive approach to investigating a person's development.

It is noteworthy that in Bronfenbrenner's theory, each system is not independent of the others; all these systems work together as a holistic, complex network to influence a child's development (e.g., Baker, Cobley, Schorer, & Wattie, 2017). A negative relationship between any of these systems can cause harm to the child. In research, it is important to separate the systems in a practical way. In reality, however, the holistic system never comes apart. The systems are always comprehensively integrated. Isolation of an individual system never permits insight into the big picture (Placek, Nishimura, Hudanick, Stephens, & Madhivanan, 2019).

Placek and colleagues (2019) point out that we need to recognize factors through the hierarchical structure of various systems in our society.

In terms of long-term care for older adults, we need to consider staff members, clients, funders, and policies (Greenfield, 2012). For example, Johri, Beland, and Bergman (2003) reviewed HCBS programs in the Organisation for Economic Co-operation and Development (OECD) countries and found that having an integrated care system—that is, combining case management, geriatric assessment, financial support policies, and a multidisciplinary team—can create efficient care to reduce the rate of hospitalization and decrease the cost of care (Johri et al., 2003). In order to develop a cost-effective system for frail older adults, all systems must be integrated to provide holistic care (Hastings, 2019; Jaffe et al., 2019). The strength of Bronfenbrenner's theory is its holistic and comprehensive approach, which is very useful for analyzing eldercare systems and services.

Bronfenbrenner's (1979) social-ecological theory, however, is not without its weaknesses. First, it takes a person-centered approach so the opportunity for organizational analysis is limited (Maes, Vanhalst, Spithoven, Van den Noortgate, & Goossens, 2016). In reality, many events occur at the organizational level that are beyond the control of the individual. Second, the theory assumes that the influence of higher layers of the environment operates according to a hierarchy, that is, the influence of a higher-level environment must pass through the lower levels before it eventually reaches the individual. In reality, though, relationships between an environment and a person can happen among any layers of the environment. For example, having decided to increase the pensions of people living in rural areas, the Chinese government directly improved the quality of their lives (Cheng, Liu, Zhang, & Zhao, 2018). In this case, the macrosystem can have a direct influence on people's lives in rural areas.

Another weakness is that individual agency is not a substantial factor in the theory because the ecological approach emphasizes the influence of the external environment on individual development. The same environment, however, does not necessarily mean that all those who are subject to it develop in the same way. For example, previous research shows that even if people live in an environment with a large supply of vegetables and public fitness, there is no guarantee that local residents will be able to use these resources (Copping & Campbell, 2015). In fact, a person has choices in his or her environment (Guay, Marsh, McInerney, & Craven, 2015; Sawitri, Hadiyanto, & Hadi, 2015). A person can also make changes to the environment. For example, some older adults do not want to interact with their neighborhood, whereas others take an active role in their surroundings (Greenfield, 2012).

In sum, Bronfenbrenner's social-ecological theory offers a broadened scope of theoretical explanations. The time frame in the theory also allows researchers to go beyond cross-sectional studies to extend their analysis to longitudinal studies and thus observe the influences of multiple levels of social systems on development in a long-term way (Hayes et al., 2017). The framework of his theory can help us to examine various interactions taking place among elderly clients and the HCBS program. It can also help us to understand the community context that the HCBS program serves and the overall connectedness of the community environment.

2.3 Greenfield's Applied Ecological Theory and Aging in Place

Dr. Emily Greenfield, an associate professor at the Rutgers School of Social Work, focuses her research on aging and services for older adults and family caregivers. In order to understand the factors that prevent or facilitate a person's choice to age in place, Greenfield pointed out that we need to study that person's comprehensive and holistic environment and relationships. She modified Bronfenbrenner's (1979) social-ecological theory to conceptualize an approach to studying aging in place (Greenfield, 2012). The new approach has six subsystems to describe the ecological model of aging in place: Environment–individual relationships, distal–proximal relationships, organizational relationships, individual agency, targeted subcultural influences, and time and life-course influences. The following sections elaborate on each one of these subsystems.

2.3.1 Environment–Individual Relationships

The concept of environment in Greenfield's applied ecological theory suggests a place in which a person lives, experiences, behaves, and perceives (Bronfenbrenner, 1979; Greenfield, 2012; Lawton, Winter, Kleban, & Ruckdeschel, 1999). This environment consists of two parts: the physical environment and the social environment.

The physical environment focuses on physical features, such as a building, an office, or a room where a person lives. It "consists of inanimate objects, space, and material structures with which an individual directly or indirectly engages (e.g., having wide door frames within one's home through which wheelchairs can pass)" (Greenfield, 2012, p. 3). Physical function interacts with and is subject to the effects of the person's physical surroundings. Moreover, physical function results from the interaction between physical-biological features and physical-environmental features (Karmiloff-Smith, 2018). For instance, Karmiloff-Smith (2018) argued that developmental disorders involve interactions between both genes and the environment.

The physical environment can indeed prevent a person from aging in place when the person is no longer physically capable of living independently within that place (Greenfield, 2012). If an older adult cannot physically overcome such environmental barriers, he or she may experience difficulties and obstacles of living in such a place. Kahana, Lovegreen, Kahana, and Kahana (2003) pointed out that older adults tend to have limited housing options when facing mobility difficulties (Kahana et al., 2003). Greenfield (2012) notes that if people with reduced mobility live on the first floor, their life will not likely change significantly but if these people live on the second floor or above without an elevator in the building, their life will be very difficult. As a result, the World Health Organization has introduced the guidelines of global age-friendly cities, requiring urban planners and policymakers to consider physical environments such as outdoor spaces, buildings, transportation, and housing to make cities age-friendly (Emlet & Moceri, 2011; Jeste et al., 2016; Klimczuk & Tomczyk, 2016; Noordzij, Beenackers, Diez Roux, & van Lenthe, 2019; Phillipson & Buffel, 2018; van Staalduinen, Bond, Dantas, & Jegundo, 2018; Webber, Porter, & Menec, 2010; World Health Organization, 2007). In fact, accessibility of resources, services, and facilities are critical dimensions of quality for community-dwelling older adults (Kahana et al., 2003). People of different ages and health conditions have unique needs that the environment must account for. For example, older adults are at risk of shrinking life space and restricted mobility in their later years (Kahana et al., 2003). In those cases, barrier-free buildings and streets can help people with disabilities and older people increase their mobility and independence (Clarke, Ailshire, Bader, Morenoff, & House, 2008; Ferris, Glicksman, & Kleban, 2014; Rosso, Auchincloss, & Michael, 2011; Sakellariou, 2015; Scheidt & Schwarz, 2013).

Since 2009, the National Aging Office of China has carried out pilot work on the "age-friendly city." This idea has been proposed to remove barriers to participation in family, community, and social life, and to form a friendly urban environment for older adults. The purpose is to develop cities that can help older people stay healthy and active (World Health Organization, 2007). In Shanghai, the government launched the age-friendly city guidelines in 2015 (Kano, Rosenberg, & Dalton, 2018). These guidelines mainly target the physical environment, including (a) green spaces and outdoor facilities that encourage people to go out and have spaces for respite; studies have reported that older people in Shanghai like relaxing rest areas, especially outdoor green spaces, in their city because they like to walk around in these areas (World Health Organization, 2007; Ying, Ning, & Xin, 2015); (b) public transportation and walking accessibility that can encourage people to go out (Li & Loo, 2016); (c) housing and security that keep older adults safe at home (Y. Liu, Dijst, Faber, Geertman, & Cui, 2017); (d) welfare programs that create a safe network for low-income older adults; (e) social services and healthcare programs that provide aging-related services for people who need them; (f) entertainment, educational, and physical exercise facilities for older adults; (g) social engagement networks that can encourage people to participate in civic activities and programs; and (h) cultural environments that improve intergenerational relationships and respect for older adults (Kano et al., 2018). These policies aim to focus on multiple levels of the environment that contribute to creating an age-friendly city (H. Chen & Adamek, 2017).

Later in 2016, the government of Shanghai concentrated on the meso-level environment and launched a "15-minute community life zone" plan (Hou & Liu, 2017). The new plan recommends that Shanghai build 15-minute comfortable living zones for community life by 2040. In terms of the service radius, most citizens should enjoy a "15-minute walking distance life zone," which integrates public services, life services, and medical services, achieving a full-coverage public services plan (Hou & Liu, 2017). Residents should have all basic services, including schools, hospitals, grocery shopping, and public activity spaces within a 15-minute walking distance from their homes. The new Shanghai policy seeks to promote a safe, friendly, and comfortable social environment for older adults (Hou & Liu, 2017).

Another essential part of the environment–individual relationship is the social environment, which reflects a broad social setting, like the macro-level environment in Bronfenbrenner's social-ecological theory (Bronfenbrenner, 1979; Greenfield, 2012). The social environment "consists of interpersonal relationships (e.g., the relationship between a care recipient and caregiver) and broader social institutions and arrangements (e.g., the provision of tax credits for family caregivers)" (Greenfield, 2012, p. 3). It emphasizes how social institutions, such as families, workplaces, organizations, and governments, influence individuals' functioning. Individual behaviors can be consistent with the norms or patterns of the groups they belong to (Farrow, Grolleau, & Ibanez, 2017; Miller & Prentice, 2016). Therefore, the social environment contains not only a macrosystem—such as culture, norms, or policies—but also interpersonal relationships that form and exist in the microsystem. Kahana and colleagues (2003) argued that environmental satisfaction and psychological well-being are based on individuals' personal preferences, needs, and social and cultural backgrounds. Kahana and Segall (1987) found that some older adults from New York did not want to move to Florida after their retirement, because they thought it was boring and lacked the stimulation they had in New York. According to the General Office of the State Council of the People's Republic of China (2019), in promoting the development of eldercare services, the society needs to build a public eldercare service system based on home-based, community-based, institutional supplemental, and medical care to ensure that everyone can enjoy basic eldercare services by 2022.

2.3.2 Distal and Proximal Environments

Greenfield identifies the relationship between local organizations and the larger environment as the relationship between the proximal and the distal, which emphasizes the interactions among institutions and the interactions between organizations and their environment (Greenfield, 2012). The distal environment represents a broader scale of environment, or a so-called upstream area (Krieger, 2008); it involves all environments in the macrosystem, such as policy, laws, cultures, norms, and social class systems (Cho, Martin, Poon, & Georgia Centenarian Study, 2015; Hawkley et al., 2008; Jacobsen et al., 2018; Krieger, 2008; Stowe & Cooney, 2015). These environments have an impact on the allocation of resources for the population. City policies on aging in place can be seen as a distal environment (Greenfield, 2012). The proximal environment, also called a downstream area (Krieger, 2008), represents institutions or organizations that people interact with daily and directly; they are physically close to individuals and provide opportunities for face-to-face contact. The downstream areas are the micro- or meso-level environments in Bronfenbrenner's social-ecological theory (1979). Proximal environments can also include a small-group environment, such as group norms (Farrow et al., 2017; Miller & Prentice, 2016). They also contain characteristics of the small environment that are relatively physically close to an individual, such as the features of the neighborhood or neighborhood relationships. These features are beyond the control of individuals. Both distal and proximal environments constitute living systems for individuals and can have an immediate impact on individuals' situations (e.g., Copping & Campbell, 2015; Furr-Holden et al., 2015; Hohl et al., 2017).

Greenfield (2012) argues that we need to understand how human activities are affected by both proximal and distal factors and how the relationship between distal and proximal environments creates significant differences among different populations. Individuals usually influence distal settings by altering elements of proximal environments, and the relationship between distal and proximal environments is dynamic (Greenfield, 2012). Identifying proximal factors is important for designing immediate intervention programs to improve people's quality of life; however, these proximal factors are determined by the distal system (Krieger, 2008). For example, if a new policy on aging allows individuals to have meals in community senior centers, theoretically, the policy can influence caregivers by decreasing their workload at home. Hong and Harrington (2016) analyzed the National Long-Term Caregiver Survey in the United States and revealed that caregiver stress is associated with fewer resources and a higher burden. Having more support from communities is associated with a lower burden and better self-perceived health among caregivers. An altered workload can also have an impact on the relationship between caregivers and care recipients (Gresham, Heffernan, & Brodaty, 2018; Hong & Harrington, 2016; Kristof, Fortinsky, Kellett, Porter, & Robison, 2017).

2.3.3 *Organizational Relationships*

Greenfield's ecological theory focuses not only on the individual-level relationships but on the networks among organizations, including cultures, norms, and values, which are very important for building an environment in which a person can thrive (Greenfield, 2012). Organizations include neighborhoods, communities, local businesses, and healthcare and social-service systems. The way organizations interact with each other can promote or harm individuals' functioning (Greenfield, 2012; Schnackenberg & Tomlinson, 2016).

For example, if a healthcare organization wants to conduct an intervention program in a community, it may need help from local groups. Community-based groups usually play a role as gatekeepers. Local people tend to know and trust the local organizations. Thus, the relationship between the intervention program and local organizations is crucial to whether the intervention can be successfully introduced; both of them need to work together to improve the overall health of the community. In such a case, the local organization becomes a stakeholder for the local community (Ahamed et al., 2008; Bovaird, 2007; Bovaird & Löffler, 2003; Galafassi et al., 2017; Gironda et al., 2010; McCrea, 2016; Schnackenberg & Tomlinson, 2016). Schnackenberg and Tomlinson (2016) argued that fostering trust between an organization and its stakeholders is very important to the success of the organization.

Another example is the latest national policy in China, "Promoting the Development of Elderly Care Services" (General Office of the State Council of the People's Republic of China, 2019), which requires that all levels of government spend no less than 55% of public welfare lottery funds on social welfare services to support the development of eldercare services. Tax and fee reduction and exemption support policies will be given to older adults' service institutions that provide daily care for seniors, rehabilitation care, food assistance, and other services in communities. Eldercare services will be included in the government's purchase service catalogue, focusing on services such as living care, rehabilitation care, institutional operations, social work, and personnel training. The project will be coordinated by the Ministry of Finance, the Ministry of Civil Affairs, the Health Committee, and local governments at all levels (General Office of the State Council of the People's Republic of China, 2019). The relationships among these departments will be very important for the implementation of this new policy.

Greenfield suggests that interactions among social systems should be a "third dimension" for understanding the relationship between individuals and the macro environment (Greenfield, 2012). Indeed, individuals are embedded within different social groups and systems with various social structures. These connected or disconnected organizations and social systems can strongly influence an individual's development (Greenfield, 2012). For example, the policy of promoting the development of eldercare services encourages primary healthcare institutions and medical staff to establish contracted service relationships with elderly families in China (General Office of the State Council of the People's Republic of China, 2019). In other words, the relationship between primary healthcare facilities and the medical staff and the relationship between medical staff and older adults' families are both critical for the quality of life among older adults in China.

2.3.4 Environment and Individual Agency

Unlike Bronfenbrenner's social-ecological theory, which tends to treat the individual as an object, Greenfield emphasizes individual agency (Greenfield, 2012). That is, human beings do not simply wait passively to be changed by their environment; they have the power to shape their environment through particular actions (Dong & Simon, 2015; Sawitri et al., 2015; Wahl, Iwarsson, & Oswald, 2012).

Older adults hold different attitudes toward support. Some may accept help from their neighbors while others may decline such help (Dong, 2017; Dong & Simon, 2015; Washington State Department of Social and Health Services, 2019). Some older adults even like to lead in informal or formal networks and offer help to other older adults instead of receiving formal support (McDonough & Davitt, 2011). Previous studies on elderly volunteers also found that they have great motivation to serve society and demonstrate their value (Allen et al., 2016; Cattan, Hogg, & Hardill, 2011; L.-K. Chen, 2016). The Chinese government's new policy promoting the development of eldercare services (General Office of the State Council of the People's Republic of China, 2019) also demonstrates the need for supporting voluntary services for older adults and actively exploring mutual assistance and eldercare services.

Individuals have the capacity to generate and shape their environments (Chatterjee, Pereira, & Bates, 2018; Rigby, Woulfin, & März, 2016). Greenfield (2012) also argued that we should be aware of how people can affect their environment by the way they evolve. For example, Kahana et al. (2003) suggested that community-based programs can request client feedback through surveys; thus, clients can be not only service recipients who benefit from a program but also active collaborators who sustain and improve the program (Greenfield, 2012). Greenfield advocated designing interventions that empower older adults to engage their own leadership, such as "leading, planning, or maintaining" a program (Greenfield, 2012; McDonough & Davitt, 2011). Programs for older adults should create partnerships with their elderly clients to nurture mutual relationships that encourage shared support. Older adults can also help to promote programs development and manage the program by themselves.

2.3.5 Environment and Subcultural Groups

Even when many people live in and interact with the same environment, they may not be influenced in the same way by that environment. Personal characteristics, such as gender, age, health status, education, and family background, can affect the relationship between individuals and their environment (Greenfield, 2012).

Research in the United States shows that the health status of different people in the same medical environment can vary greatly, and the benefits they receive are varied as well (LaVeist, 1993; Nelson, 1992; Tamayo-Sarver, Hinze, Cydulka, & Baker, 2003; Tammemagi, Nerenz, Neslund-Dudas, Feldkamp, & Nathanson, 2005). For example, the proportion of elderly black males who have access to or make use of cancer screening is relatively small (Fillenbaum, Burchett, Kuchibhatla, Cohen, & Blazer, 2007). For older adults who are entitled to receive Medicaid support for home- and community-based services, their quality of life is determined by the state and the specific waiver program that they can have locally (Garza, 2018). In China, the latest policy promoting the development of eldercare services pointed out that future government actions need to focus on providing free or low-cost childcare services for older adults with financial difficulties, dementia, and cognitive impairment, and those without adult children (General Office of the State Council of the People's Republic of China, 2019).

Another example that is relevant to subgroup characteristics is older adults' self-perceived successful aging (Greenfield, 2012). Dannefer (2008) highlighted subgroup characteristics, arguing that older adults cannot be treated as a single group. Intra-group differences need to be factored into a comprehensive view of the aging population (Dannefer, 2008). Previous studies have shown that older adults with different cultures, health situations, gender identifications, and sexual orientations have different views on the concept of successful aging (L. Chen et al., 2019; Fabbre, 2015; Fredriksen-Goldsen, Kim, Shiu, Goldsen, & Emlet,

2015; Ji, Ling, & McCarthy, 2015; Katz & Calasanti, 2015; Martin et al., 2015; Martinson & Berridge, 2015; Peterson & Martin, 2015; Rubinstein & de Medeiros, 2015).

2.3.6 Time and Life-Course Influence

One of the main hypotheses of Bronfenbrenner's social-ecological theory (1979) is that long-term interactions are carried out between individuals and their environment. Similar to Bronfenbrenner, Greenfield (2012) highlighted the impact of previous behaviors and life events on later life. When a life event happens, it has an influence on the success of later life transitions or events. In other words, the impact of a shift in conditions on individual human development depends on its timing compared with other events in a person's life (Dannefer, 2003a, 2003b; Elder, Johnson, & Crosnoe, 2003; Elder & Shanahan, 2007).

Elder et al. (2003) pointed out that transitions often require personal and social adjustments to a life event, thus opening up possibilities for identity adjustment and behavioral change. Greenfield (2012) noted that if an older adult experiences a sudden health crisis (e.g., stroke, heart attack) and is treated in the hospital, he or she needs to gain access to a nursing home or rehabilitation center after leaving the hospital. The patient and his or her family need to adjust their lives according to such a change. Individual adjustment happens when life transitions evolve from one healthcare situation to another (Greenfield, 2012).

As new technology has developed, it has also been applied to services for older adults. China's Ministry of Industry and Information Technology, Ministry of Civil Affairs, Development and Reform Commission, and Health Commission are jointly responsible for the integration of eldercare services with internet technology in the future (General Office of the State Council of the People's Republic of China, 2019). The new policy ("Promoting the Development of Elderly Care Services") indicates that society needs to promote the development of a smart and healthy pension system and expand the application of information technology (e.g., artificial intelligence, cloud computing, robots, big data, etc.) in the field of eldercare, none of which have been used in the past.

Like Bronfenbrenner's ecological theory, Greenfield's ecological framework can also incorporate other theories and models. Her theory offers a comprehensive approach to designing research and interventions. Considering the differences among older adults, Greenfield highlights the importance of human agency in the relationship between individuals and their environment. Her theory also helps us to look beyond individuals and cast our attention toward organizations, environments, and relationships, employing multiple levels of interaction to observe individuals and organizations so that researchers can fully capture factors that essentially affect behaviors.

2.4 Chapter Summary

This chapter introduces socio-ecological theory to theorize the concept of community care ecology. Bronfenbrenner's social-ecological theory provides a solid foundation for examining the effects of multiple levels of social systems on HCBS (Bronfenbrenner, 1979; Hayes et al., 2017). Greenfield's ecological framework offers an evolving ecological perspective to understand older adults' agency and organizations in the context of community care ecology. In the multilayered community care ecology, it is essential to incorporate relevant theories and models (Greenfield, 2012). In the following chapters, we will introduce the Shanghai HCBS program and factors from each level of the environment that impact the program using socio-ecological theory.

References

Ahamed, T., Khan, M. I. N., Takigawa, T., Koike, M., Tasnim, F., & Zaman, J. M. Q. (2008). Resource management for sustainable development: A community- and GIS-based approach. *Environment, Development and Sustainability, 11*(5), 933–954. https://doi.org/10.1007/s10668-008-9159-y.

Allen, R. S., Azuero, C. B., Csikai, E. L., Parmelee, P. A., Shin, H. J., Kvale, E., ... Burgio, L. D. (2016). "It was very rewarding for me ...": Senior volunteers' experiences with implementing a reminiscence and creative activity intervention. *The Gerontologist*, *56*(2), 357–367. https://doi.org/10.1093/geront/gnu167.

Baker, J., Cobley, S., Schorer, J., & Wattie, N. (2017). *Routledge handbook of talent identification and development in sport*. New York: Taylor & Francis.

Bandura, A. (2009). Social cognitive theory of mass communication (Chapter 6). In J. Bryant & M. B. Oliver (Eds.), Media effects: Advances in theory and research (pp. 110–140). Routledge. https://doi.org/10.4324/9780203877111-12

Barrows, S. P. (2016). China's one-child policy. *JAMA, 315*(21), 2349–2350. https://doi.org/10.1001/jama.2016.2192.

Bovaird, T. (2007). Beyond engagement and participation: User and community coproduction of public services. *Public Administration Review, 67*(5), 846–860. https://doi.org/10.1111/j.1540-6210.2007.00773.x.

Bovaird, T., & Löffler, E. (2003). Evaluating the quality of public governance: Indicators, models and methodologies. *International Review of Administrative Sciences, 69*(3), 313–328. https://doi.org/10.1177/0020852303693002.

Bowlby, S., & McKie, L. (2019). Care and caring: An ecological framework. *Area, 51*(3), 532–539. https://doi.org/10.1111/area.12511.

Bronfenbrenner, U. (1977). Toward an experimental ecology of human development. *American Psychologist, 32*(7), 513–531. https://doi.org/10.1037/0003-066X.32.7.513.

Bronfenbrenner, U. (1979). *The ecology of human development: Experiments by nature and design*. Cambridge: Harvard University Press.

Bronfenbrenner, U. (2001). The bioecological theory of human development. In N. J. Smelser & B. Baltes (Eds.), *International encyclopedia of the social and behavioral sciences*. Amsterdam: Elsevier.

Bronfenbrenner, U., & Morris, P. A. (2007). The bioecological model of human development (Chapter 14). In W. Damon & R. M. Lerner (Eds.), *Handbook of child psychology, theoretical models of human development* (pp. 793–828). Hoboken: Wiley.

Carstensen, L. L. (1992). Social and emotional patterns in adulthood: Support for socioemotional selectivity theory. *Psychology and Aging, 7*(3), 331–338.

Carstensen, L. L. (1995). Evidence for a life-span theory of socioemotional selectivity. *Current Directions in Psychological Science, 4*(5), 151–156. https://doi.org/10.1111/1467-8721.ep11512261.

Carstensen, L. L., Isaacowitz, D. M., & Charles, S. T. (1999). Taking time seriously: A theory of socioemotional selectivity. *American Psychologist, 54*(3), 165–181. https://doi.org/10.1037/0003-066X.54.3.165.

Casper, W. J., Vaziri, H., Wayne, J. H., DeHauw, S., & Greenhaus, J. (2018). The jingle-jangle of work–nonwork balance: A comprehensive and meta-analytic review of its meaning and measurement. *Journal of Applied Psychology, 103*(2), 182–214. https://doi.org/10.1037/apl0000259.

Cattan, M., Hogg, E., & Hardill, I. (2011). Improving quality of life in ageing populations: What can volunteering do? *Maturitas, 70*(4), 328–332. https://doi.org/10.1016/j.maturitas.2011.08.010.

Chan, R. K. H. (2006). Risk and its management in post-financial crisis Hong Kong. *Social Policy & Administration, 40*(2), 215–229. https://doi.org/10.1111/j.1467-9515.2006.00485.x.

Chatterjee, A., Pereira, A., & Bates, R. (2018). Impact of individual perception of organizational culture on the learning transfer environment. *International Journal of Training and Development, 22*(1), 15–33. https://doi.org/10.1111/ijtd.12116.

Chen, H., & Adamek, M. (2017). Civic engagement of older adults in mainland China: Past, present, and future. *The International Journal of Aging and Human Development, 85*(2), 204–226. https://doi.org/10.1177/0091415016685326.

Chen, L., & Ye, M. (2013). The role of children's support in elders' decisions to live in a yanglaoyuan (residential long-term care): Children's support and decision to live in a yanglaoyuan. *Journal of Cross-Cultural Gerontology, 28*(1), 75–87. https://doi.org/10.1007/s10823-012-9185-y.

Chen, L., Ye, M., & Kahana, E. (2019). A self-reliant umbrella: Defining successful aging among the old-old (80+) in Shanghai. *Journal of Applied Gerontology*, 0733464819842500. https://doi.org/10.1177/0733464819842500.

Chen, L.-K. (2016). Not just helping: What and how older men learn when they volunteer. *Educational Gerontology, 42*(3), 175–185. https://doi.org/10.1080/03601277.2015.1085786.

Chen, X. (2015). Old age pension and intergenerational living arrangements: A regression discontinuity design. *Review of Economics of the Household*, 1–22. https://doi.org/10.1007/s11150-015-9304-y.

Cheng, L., Liu, H., Zhang, Y., & Zhao, Z. (2018). The health implications of social pensions: Evidence from China's new rural pension scheme. *Journal of Comparative Economics, 46*(1), 53–77. https://doi.org/10.1016/j.jce.2016.12.002.

China Daily. (2018, June 23). *China's senior citizens flock back to school*. https://www.chinadaily.com.cn/a/201806/23/WS5b2d4787a3103349141de584.html.

China Daily. (2019, October 9). *Ministry transfers assets worth $21 bln to national social security fund*. http://www.ecns.cn/news/2019-10-09/detail-ifzpquht1928069.shtml.

Cho, J., Martin, P., Poon, L. W., & Georgia Centenarian Study. (2015). Successful aging and subjective well-being among oldest-old adults. *The Gerontologist, 55*(1), 132–143. https://doi.org/10.1093/geront/gnu074.

Choi, M., & Fong, D. (2017). Investigation on Chinese senior tourists. *Journal of Tourism & Hospitality, 06*(05). https://doi.org/10.4172/2167-0269.1000307

Clarke, P., Ailshire, J. A., Bader, M., Morenoff, J. D., & House, J. S. (2008). Mobility disability and the urban built environment. *American Journal of Epidemiology, 168*(5), 506–513. https://doi.org/10.1093/aje/kwn185.

Clausen, J. A. (1993). Introduction: Adolescent planful competence and the shaping of the life course (Chapter 1). *American lives: Looking back at the children of the great depression* (pp. 1–25). New York: The Free Press.

Copping, L. T., & Campbell, A. (2015). The environment and life history strategies: Neighborhood and individual-level models. *Evolution and Human Behavior, 36*(3), 182–190. https://doi.org/10.1016/j.evolhumbehav.2014.10.005.

Crosnoe, R. (2004). Social capital and the interplay of families and schools. *Journal of Marriage and Family, 66*(2), 267–280. https://doi.org/10.1111/j.1741-3737.2004.00019.x.

Dannefer, D. (2003a). Cumulative advantage/disadvantage and the life course: Cross-fertilizing age and social science theory. *The Journals of Gerontology Series B: Psychological Sciences and Social Sciences, 58*(6), S327–S337. https://doi.org/10.1093/geronb/58.6.S327.

Dannefer, D. (2003b). Toward a global geography of the life course. In J. T. Mortimer & M. J. Shanahan (Eds.), *Handbook of the life course* (pp. 647–659). Springer US. http://link.springer.com/chapter/10.1007/978-0-306-48247-2_29.

Dannefer, D. (2008). The waters we swim: Everyday social processes, macrostructural realities, and human aging. In K. W. Schaie & R. P. Abeles, *Social structures and aging individuals: Continuing challenges* (pp. 3–18). Springer. http://public.eblib.com/choice/publicfullrecord.aspx?p=423300.

Dong, X. (2017). Elder self-neglect: Research and practice. *Clinical Interventions in Aging, 12,* 949–954. https://doi.org/10.2147/CIA.S103359.

Dong, X., & Simon, M. A. (2015). Elder self-neglect is associated with an increased rate of 30-day hospital readmission: Findings from the Chicago health and aging project. *Gerontology, 61*(1), 41–50. https://doi.org/10.1159/000360698.

Dong, X., Zhang, M., & Simon, M. A. (2014). The expectation and perceived receipt of filial piety among Chinese older adults in the greater Chicago area. *Journal of Aging and Health, 26*(7), 1225–1247. https://doi.org/10.1177/0898264314541697.

Elder, G. H., Jr. (1994). Time, human agency, and social change: Perspectives on the life course. *Social Psychology Quarterly, 57*(1), 4–15.

Elder, G. H., Jr., Johnson, M. K., & Crosnoe, R. (2003). The emergence and development of life course theory. In J. T. Mortimer & M. J. Shanahan (Eds.), *Handbook of the life course* (pp. 3–16). Berlin: Kluwer Academic/Pleum Publishers.

Elder, G. H., Jr., & Shanahan, M. J. (2007). The life course and human development. In *Handbook of child psychology*. Wiley. http://onlinelibrary.wiley.com/doi/10.1002/9780470147658.chpsy0112/abstract

Emlet, C. A., & Moceri, J. T. (2011). The importance of social connectedness in building age-friendly communities. *Journal of Aging Research, 2012*. https://doi.org/10.1155/2012/173247.

Espelage, D. L., & Swearer, S. M. (Eds.). (2010). A social-ecological model for bullying prevention and intervention: Understanding the impact of adults in the social ecology of youngsters. In *Handbook of bullying in schools: An international perspective* (pp. 61–72). Routledge/Taylor & Francis Group. https://doi.org/10.4324/9780203842898.

Ettner, S. L., & Grzywacz, J. G. (2001). Workers' perceptions of how jobs affect health: A social ecological perspective. *Journal of Occupational Health Psychology, 6*(2), 101–113. https://doi.org/10.1037/1076-8998.6.2.101.

Fabbre, V. D. (2015). Gender transitions in later life: A queer perspective on successful aging. *The Gerontologist, 55*(1), 144–153. https://doi.org/10.1093/geront/gnu079.

Farrow, K., Grolleau, G., & Ibanez, L. (2017). Social norms and pro-environmental behavior: A review of the evidence. *Ecological Economics, 140*, 1–13. https://doi.org/10.1016/j.ecolecon.2017.04.017.

Fei, H. T. (1939). *Peasant life in China*. Hong Kong: Hesperides Press.

Ferguson, K. T., & Evans, G. W. (2019). Social ecological theory: Family systems and family psychology in bioecological and bioecocultural perspective. In *APA handbook of contemporary family psychology: Foundations, methods, and contemporary issues across the lifespan, Vol. 1* (pp. 143–161). American Psychological Association. https://doi.org/10.1037/0000099-009.

Ferris, R. E., Glicksman, A., & Kleban, M. H. (2014). Environmental predictors of unmet home-and community-based service needs of older adults. *Journal of Applied Gerontology*. https://doi.org/10.1177/0733464814525504.

Fillenbaum, G. G., Burchett, B. M., Kuchibhatla, M. N., Cohen, H. J., & Blazer, D. G. (2007). Effect of cancer screening and desirable health behaviors on functional status, self-rated health, health service use and mortality. *Journal of the American Geriatrics Society, 55*(1), 66–74. https://doi.org/10.1111/j.1532-5415.2006.01009.x.

Fredriksen-Goldsen, K. I., Kim, H.-J., Shiu, C., Goldsen, J., & Emlet, C. A. (2015). Successful aging among LGBT older adults: Physical and mental health-related quality of life by age group. *The Gerontologist, 55*(1), 154–168. https://doi.org/10.1093/geront/gnu081.

Furr-Holden, C. D. M., Lee, M. H., Johnson, R., Milam, A. J., Duncan, A., Reboussin, B. A., Leaf, P. J., & Ialongo, N. S. (2015). Neighborhood environment and marijuana use in urban young adults. *Prevention Science, 16*(2), 268–278. https://doi.org/10.1007/s11121-014-0497-8.

Galafassi, D., Daw, T. M., Munyi, L., Brown, K., Barnaud, C., & Fazey, I. (2017). Learning about social-ecological trade-offs. *Ecology and Society, 22*(1), 2. https://doi.org/10.5751/ES-08920-220102.

Garza, L. (2018). *Understanding barriers affecting elders' low utilization of HCBS: A multiple case study of HCBS social workers* (Dissertations). Luzmaria.

General Office of the State Council of the People's Republic of China. (2019, April 16). *Promoting the development of elderly care services opinions*. http://www.gov.cn/zhengce/content/2019-04/16/content_5383270.htm.

George, L. K. (1993). Sociological perspectives on life transitions. *Annual Review of Sociology, 19*, 353–373.

Gironda, M. W., Lefever, K., Delagrammatikas, L., Nerenberg, L., Roth, R., Chen, E. A., & Northington, K. R. (2010). Education and training of mandated reporters: Innovative models, overcoming challenges, and lessons learned. *Journal of Elder Abuse & Neglect, 22*(3–4), 340–364. https://doi.org/10.1080/08946566.2010.490188.

Glanz, K., Rimer, B. K., & Viswanath, K. (2015). *Health behavior: Theory, research, and practice*. New York: Wiley.

Görg, C., Brand, U., Haberl, H., Hummel, D., Jahn, T., & Liehr, S. (2017). Challenges for social-ecological transformations: Contributions from social and political ecology. *Sustainability, 9*(7), 1045. https://doi.org/10.3390/su9071045.

Greenfield, E. A. (2012). Using ecological frameworks to advance a field of research, practice, and policy on aging-in-place initiatives. *The Gerontologist, 52*(1), 1–12. https://doi.org/10.1093/geront/gnr108.

Gresham, M., Heffernan, M., & Brodaty, H. (2018). The going to stay at home program: Combining dementia caregiver training and residential respite care. *International Psychogeriatrics, 30*(11), 1697–1706. https://doi.org/10.1017/S1041610218000686.

Guay, F., Marsh, H., McInerney, D. M., & Craven, R. G. (2015). *Selfconcept, motivation and identity: Underpinning success with research and practice*. IAP.

Guo, M., Xu, L., Liu, J., Mao, W., & Chi, I. (2016). Parent–child relationships among older Chinese immigrants: The influence of co-residence, frequent contact, intergenerational support and sense of children's deference. *Ageing & Society, 36*(07), 1459–1482. https://doi.org/10.1017/S0144686X15000446.

Halldorsdottir, T., & Binder, E. B. (2017). Gene × environment interactions: From molecular mechanisms to behavior. *Annual Review of Psychology, 68*(1), 215–241. https://doi.org/10.1146/annurev-psych-010416-044053.

Hastings, A. C. (2019). *Health for the whole person: The complete guide to holistic medicine*. Abingdon: Routledge.

Hawkley, L. C., Hughes, M. E., Waite, L. J., Masi, C. M., Thisted, R. A., & Cacioppo, J. T. (2008). From social structural factors to perceptions of relationship quality and loneliness: The Chicago health, aging, and social relations study. *The Journals of Gerontology Series B: Psychological Sciences and Social Sciences, 63*(6), S375–S384.

Hayes, N., O'Toole, L., & Halpenny, A. M. (2017). *Introducing Bronfenbrenner: A guide for practitioners and students in early years education*. Abingdon: Routledge.

He, G., Xie, J., Zhou, J., Zhong, Z., Qin, C., & Ding, S. (2016). Depression in left-behind elderly in rural China: Prevalence and associated factors. *Geriatrics & Gerontology International, 16*(5), 638–643. https://doi.org/10.1111/ggi.12518.

Hickerson, B., Moore, A., Oakleaf, L., Edwards, M., James, P. A., Swanson, J., & Henderson, K. A. (2008). The role of a senior center in promoting physical activity for older adults. *Journal of Park and Recreation Administration, 26*(1). https://js.sagamorepub.com/jpra/article/view/1332.

Hohl, B. C., Wiley, S., Wiebe, D. J., Culyba, A. J., Drake, R., & Branas, C. C. (2017). Association of drug and alcohol use with adolescent firearm homicide at individual, family, and neighborhood levels. *JAMA Internal Medicine, 177*(3), 317–324. https://doi.org/10.1001/jamainternmed.2016.8180.

Holden, G. W. (2019). *Parents and the dynamics of child rearing*. Abingdon: Routledge.

Hong, M., & Harrington, D. (2016). The effects of caregiving resources on perceived health among caregivers. *Health and Social Work, 41*(3), 155–163. https://doi.org/10.1093/hsw/hlw025.

Hou, L., & Liu, Y. (2017). Life circle construction in China under the idea of collaborative governance: A comparative study of Beijing, Shanghai and Guangzhou. *Geographical Review of Japan Series B, 90*(1), 2–16. https://doi.org/10.4157/geogrevjapanb.90.2.

Jacobsen, P. B., DeRosa, A. P., Henderson, T. O., Mayer, D. K., Moskowitz, C. S., Paskett, E. D., & Rowland, J. H. (2018). Systematic review of the impact of cancer survivorship care plans on health outcomes and health care delivery. *Journal of Clinical Oncology, 36*(20), 2088–2100. https://doi.org/10.1200/JCO.2018.77.7482.

Jaffe, R. C., Bergin, C. R., Loo, L. K., Singh, S., Uthlaut, B., Glod, S. A., … Gentilesco, B. (2019). Reactive, holistic, proactive: Practical applications of the AAIM learning and working environment conceptual model. *The American Journal of Medicine, 132*(8), 995–1000. https://doi.org/10.1016/j.amjmed.2019.04.004.

Jeste, D. V., Blazer, D. G., Buckwalter, K. C., Cassidy, K.-L. K., Fishman, L., Gwyther, L. P., … Feather, J. (2016). Age-friendly communities initiative: Public health approach to promoting successful aging. *The American Journal of Geriatric Psychiatry, 24*(12), 1158–1170. https://doi.org/10.1016/j.jagp.2016.07.021.

Ji, H., Ling, J., & McCarthy, V. L. (2015). Successful aging in the United States and China: A theoretical basis to guide nursing research, practice, and policy. *Journal of Transcultural Nursing, 26*(2), 129–136. https://doi.org/10.1177/1043659614526257.

Johri, M., Beland, F., & Bergman, H. (2003). International experiments in integrated care for the elderly: A synthesis of the evidence. *International Journal of Geriatric Psychiatry, 18*(3), 222–235. https://doi.org/10.1002/gps.819.

Kahana, E., Lee, J. E., Kahana, J., Goler, T., Kahana, B., Shick, … Barnes, K. (2015). Childhood autism and proactive family coping: Intergenerational perspectives. *Journal of Intergenerational Relationships, 13*(2), 150–166. https://doi.org/10.1080/15350770.2015.1026759.

Kahana, E., Lovegreen, L., Kahana, B., & Kahana, M. (2003). Person, environment, and person-environment fit as influences on residential satisfaction of elders. *Environment and Behavior, 35*(3), 434–453. https://doi.org/10.1177/0013916503035003007.

Kahana, E., & Segall, M. (1987). Deinstitutionalization. In G. Maddox (Ed.), *Encyclopedia of aging* (pp. 163–164). Springer. https://www.researchgate.net/publication/278677123_Deinstitutionalization.

Kahana, E., & Young, R. (1990). Clarifying the caregiving paradigm: Challenges for the future. In D. E. Blegel & A. Blum (Eds.), *Aging and caregiving: Theory, research and practice* (pp. 76–97). Thousand Oaks: Sage.

Kano, M., Rosenberg, P. E., & Dalton, S. D. (2018). A global pilot study of age-friendly city indicators. *Social Indicators Research, 138*(3), 1205–1227. https://doi.org/10.1007/s11205-017-1680-7.

Karmiloff-Smith, A. (2018). *Thinking developmentally from constructivism to neoconstructivism: The selected works of Annette Karmiloff-Smith*. Abingdon: Routledge.

Katz, S., & Calasanti, T. (2015). Critical perspectives on successful aging: Does it "appeal more than it illuminates"? *The Gerontologist, 55*(1), 26–33. https://doi.org/10.1093/geront/gnu027.

Klimczuk, A., & Tomczyk, Ł. (2016). Smart, age-friendly cities and communities: The emergence of socio-technological solutions in the Central and Eastern Europe. *Active and Assisted Living: Technologies and Applications*, 335–359. https://doi.org/10.1049/PBHE006E_ch17.

Kohn, A. (1999). *Punished by rewards: The trouble with gold stars, incentive plans, a's, praise, and other bribes* (2nd ed.). Boston: Mariner Books.

Krieger, N. (2008). Proximal, distal, and the politics of causation: What's level got to do with it? *American Journal of Public Health, 98*(2), 221–230. https://doi.org/10.2105/AJPH.2007.111278.

Kristof, L., Fortinsky, R. H., Kellett, K., Porter, M., & Robison, J. (2017). Experiences of informal caregivers of older adults transitioned from nursing homes to the community through the money follows the person demonstration. *Journal of Aging & Social Policy, 29*(1), 20–34. https://doi.org/10.1080/08959420.2016.1187034.

Langille, J.-L. D., & Rodgers, W. M. (2010). Exploring the influence of a social ecological model on school-based physical activity. *Health Education & Behavior, 37*(6), 879–894. https://doi.org/10.1177/1090198110367877.

LaVeist, T. A. (1993). Segregation, poverty, and empowerment: Health consequences for African Americans. *The Milbank Quarterly, 71*(1), 41–64. https://doi.org/10.2307/3350274.

Lawton, M. P., Winter, L., Kleban, M. H., & Ruckdeschel, K. (1999). Affect and quality of life: Objective and subjective. *Journal of Aging and Health, 11*(2), 169–198. https://doi.org/10.1177/089826439901100203.

Lee, G. R., & DeMaris, A. (2007). Widowhood, gender, and depression a longitudinal analysis. *Research on Aging, 29*(1), 56–72. https://doi.org/10.1177/0164027506294098.

Li, L., & Loo, B. P. Y. (2016). Towards people-centered integrated transport: A case study of Shanghai hongqiao comprehensive transport hub. *Cities, 58*, 50–58. https://doi.org/10.1016/j.cities.2016.05.003.

Liu, T., & Sun, L. (2016). Pension reform in China. *Journal of Aging & Social Policy, 28*(1), 15–28. https://doi.org/10.1080/08959420.2016.1111725.

Liu, Y., Dijst, M., Faber, J., Geertman, S., & Cui, C. (2017). Healthy urban living: Residential environment and health of older adults in Shanghai. *Health & Place, 47*, 80–89. https://doi.org/10.1016/j.healthplace.2017.07.007.

Ma, P. H. X., Chan, Z. C. Y., & Loke, A. Y. (2017). The socio-ecological model approach to understanding barriers and facilitators to the accessing of health services by sex workers: A systematic review. *AIDS and Behavior, 21*(8), 2412–2438. https://doi.org/10.1007/s10461-017-1818-2.

Maes, M., Vanhalst, J., Spithoven, A. W. M., Van den Noortgate, W., & Goossens, L. (2016). Loneliness and attitudes toward aloneness in adolescence: A person-centered approach. *Journal of Youth and Adolescence, 45*(3), 547–567. https://doi.org/10.1007/s10964-015-0354-5.

Martin, P., Kelly, N., Kahana, B., Kahana, E., Willcox, B. J., Willcox, D. C., & Poon, L. W. (2015). Defining successful aging: A tangible or elusive concept? *The Gerontologist*, *55*(1), 14–25. https://doi.org/10.1093/geront/gnu044.

Martinson, M., & Berridge, C. (2015). Successful aging and its discontents: A systematic review of the social gerontology literature. *The Gerontologist, 55*(1), 58–69. https://doi.org/10.1093/geront/gnu037.

McCrea, D. M. (2016). *Home- and community-based long-term services and supports: An analysis of provisions contained in the Affordable Care Act* (M.S., Utica College). http://search.proquest.com/docview/1835846590/abstract/E3396779D95B41EBPQ/1.

McDonough, K. E., & Davitt, J. K. (2011). It takes a village: Community practice, social work, and aging-in-place. *Journal of Gerontological Social Work, 54*(5), 528–541. https://doi.org/10.1080/01634372.2011.581744.

Miller, D. T., & Prentice, D. A. (2016). Changing norms to change behavior. *Annual Review of Psychology, 67*(1), 339–361. https://doi.org/10.1146/annurev-psych-010814-015013.

Moran, M. B., Frank, L. B., Zhao, N., Gonzalez, C., Thainiyom, P., Murphy, S. T., & Ball-Rokeach, S. J. (2016). An argument for ecological research and intervention in health communication. *Journal of Health Communication, 21*(2), 135–138. https://doi.org/10.1080/10810730.2015.1128021.

Nelson, M. D. (1992). Socioeconomic status and childhood mortality in North Carolina. *American Journal of Public Health, 82*(8), 1131–1133. https://doi.org/10.2105/AJPH.82.8.1131.

Noordzij, J. M., Beenackers, M. A., Diez Roux, A. V., & van Lenthe, F. J. (2019). Age-friendly cities: Challenges for future research. *Bulletin of the World Health Organization, 97*(6), 436–437. https://doi.org/10.2471/BLT.18.224865.

Ostrom, E. (2014). Do institutions for collective action evolve? *Journal of Bioeconomics, 16*(1), 3–30. https://doi.org/10.1007/s10818-013-9154-8.

Parrott, L., & Meyer, W. S. (2012). Future landscapes: Managing within complexity. *Frontiers in Ecology and the Environment, 10*(7), 382–389. https://doi.org/10.1890/110082.

Peterson, N. M., & Martin, P. (2015). Tracing the origins of success: Implications for successful aging. *The Gerontologist, 55*(1), 5–13. https://doi.org/10.1093/geront/gnu054.

Phillipson, C., & Buffel, T. (2018). Developing age-friendly cities and communities: New directions for research and policy. *Innovation in Aging, 2*(Suppl. 1), 217–217. https://doi.org/10.1093/geroni/igy023.798.

Placek, C. D., Nishimura, H., Hudanick, N., Stephens, D., & Madhivanan, P. (2019). Reframing HIV stigma and fear: Considerations from social-ecological and evolutionary theories of reproduction. *Human Nature, 30*(1), 1–22. https://doi.org/10.1007/s12110-018-09335-z.

Posner, M. I., & Rothbart, M. K. (2018). Parenting and human brain development. In M. R. Sanders & A. Morawska (Eds.), *Handbook of parenting and child development across the lifespan* (pp. 173–199). Springer. https://doi.org/10.1007/978-3-319-94598-9_8

Powell, M. (1990). Residential environment and self-directedness among older people. *American Psychologist, 45*(5), 638–640. https://doi.org/10.1037/0003-066X.45.5.638.

Reifsnider, E., Gallagher, M., & Forgione, B. (2005). Using ecological models in research on health disparities. *Journal of Professional Nursing, 21*(4), 216–222. https://doi.org/10.1016/j.profnurs.2005.05.006.

Rigby, J. G., Woulfin, S. L., & März, V. (2016). Understanding how structure and agency influence education policy implementation and organizational change. *American Journal of Education, 122*(3), 295–302. https://doi.org/10.1086/685849.

Roberts, A. R., & Ishler, K. J. (2018). Family involvement in the nursing home and perceived resident quality of life. *The Gerontologist, 58*(6), 1033–1043. https://doi.org/10.1093/geront/gnx108.

Roberts, T., & Bowers, B. (2015). How nursing home residents develop relationships with peers and staff: A grounded theory study. *International Journal of Nursing Studies, 52*(1), 57–67. https://doi.org/10.1016/j.ijnurstu.2014.07.008.

Rosso, A. L., Auchincloss, A. H., & Michael, Y. L. (2011). The urban built environment and mobility in older adults: A comprehensive review. *Journal of Aging Research, 2011.* https://doi.org/10.4061/2011/816106.

Rubinstein, R. L., & de Medeiros, K. (2015). "Successful aging", gerontological theory and neoliberalism: A qualitative critique. *The Gerontologist, 55*(1), 34–42. https://doi.org/10.1093/geront/gnu080.

Sakellariou, D. (2015). Home modifications and ways of living well. *Medical Anthropology, 34*(5), 456–469. https://doi.org/10.1080/01459740.2015.1012614.

Sawitri, D. R., Hadiyanto, H., & Hadi, S. P. (2015). Pro-environmental behavior from a social cognitive theory perspective. *Procedia Environmental Sciences, 23,* 27–33. https://doi.org/10.1016/j.proenv.2015.01.005.

Scheidt, R. J., & Schwarz, B. (2013). *Environmental gerontology: What now?*. Abingdon: Routledge.

Schlechter, C. R. (2015). *Physical activity during recreational youth sport: Does coach training have an influence?* (Thesis, Kansas State University). https://krex.k-state.edu/dspace/handle/2097/20381

Schlechter, C. R. (2018). *The social contexts of youth settings for influencing physical activity.* Manhattan: Kansas State University.

Schnackenberg, A. K., & Tomlinson, E. C. (2016). Organizational transparency: A new perspective on managing trust in organization-stakeholder relationships. *Journal of Management, 42*(7), 1784–1810. https://doi.org/10.1177/0149206314525202.

Schunk, D. H. (2012). *Learning theories: An educational perspective* (6th ed.). London: Pearson.

Shelton, L. G. (2018). *The Bronfenbrenner primer: A guide to develecology.* Abingdon: Routledge.

Simon, M. A., Chen, R., Chang, E.-S., & Dong, X. (2014). The association between filial piety and suicidal ideation: Findings from a community-dwelling Chinese aging population. *The Journals of Gerontology Series a: Biological Sciences and Medical Sciences, 69*(Suppl. 2), S90–S97. https://doi.org/10.1093/gerona/glu142.

Stiles, J., Brown, T. T., Haist, F., & Jernigan, T. L. (2015). Brain and cognitive development. In *Handbook of child psychology and developmental science* (pp. 1–54). American Cancer Society. https://doi.org/10.1002/9781118963418.childpsy202

Stowe, J. D., & Cooney, T. M. (2015). Examining Rowe and Kahn's concept of successful aging: Importance of taking a life course perspective. *The Gerontologist, 55*(1), 43–50. https://doi.org/10.1093/geront/gnu055.

Sun, W., Aodeng, S., Tanimoto, Y., Watanabe, M., Han, J., Wang, B., ... Kono, K. (2015). Quality of life (QOL) of the community-dwelling elderly and associated factors: A population-based study in urban areas of China. *Archives of Gerontology and Geriatrics, 60*(2), 311–316. https://doi.org/10.1016/j.archger.2014.12.002.

Sun, Y. (2017). Among a hundred good virtues, filial piety is the first: Contemporary moral discourses on filial piety in urban China. *Anthropological Quarterly, 90*(3), 771–799. https://doi.org/10.1353/anq.2017.0043.

Surr, C. A., Smith, S. J., Crossland, J., & Robins, J. (2016). Impact of a person-centred dementia care training programme on hospital staff attitudes, role efficacy and perceptions of caring for people with dementia: A repeated measures study. *International Journal of Nursing Studies, 53,* 144–151. https://doi.org/10.1016/j.ijnurstu.2015.09.009.

Tamayo-Sarver, J. H., Hinze, S. W., Cydulka, R. K., & Baker, D. W. (2003). Racial and ethnic disparities in emergency department analgesic prescription. *American Journal of Public Health, 93*(12), 2067–2073. https://doi.org/10.2105/AJPH.93.12.2067.

Tammemagi, C. M., Nerenz, D., Neslund-Dudas, C., Feldkamp, C., & Nathanson, D. (2005). Comorbidity and survival disparities among black and white patients with breast cancer. *JAMA, 294*(14), 1765–1772. https://doi.org/10.1001/jama.294.14.1765.

The Standing Committee of the National People's Congress. (2019). *Law of the People's Republic of China on protection of the rights and interests of the elderly*. http://www.china.org.cn/government/laws/2007-04/17/content_1207404.htm.

Townsend, D., Taylor, L. K., Furey, A., Merrilees, C. E., Goeke-Morey, M. C., Shirlow, P., & Cummings, E. M. (2016). Measuring the macrosystem in postaccord Northern Ireland: A social–ecological approach. *Peace and Conflict: Journal of Peace Psychology, 22*(3), 282–286. https://doi.org/10.1037/pac0000185.

Tudge, J. R. H., Payir, A., Merçon-Vargas, E., Cao, H., Liang, Y., Li, J., & O'Brien, L. (2016). Still misused after all these years? A reevaluation of the uses of bronfenbrenner's bioecological theory of human development. *Journal of Family Theory & Review, 8*(4), 427–445. https://doi.org/10.1111/jftr.12165.

Turner, M. (2019). *Investigating attrition among special educators in relation to Bronfenbrenner's ecological theory* (Walden Dissertations and Doctoral Studies). https://scholarworks.waldenu.edu/dissertations/6747.

Vaccaro, A., & Kimball, E. (2019). Navigating disability in campus housing: An ecological analysis of student affairs work. *Journal of Student Affairs Research and Practice, 56*(2), 168–180. https://doi.org/10.1080/19496591.2018.1490307.

van Staalduinen, W., Bond, R., Dantas, C., & Jegundo, A. L. (2018). *Smart age-friendly cities age-friendly smart cities*. Retrieved in European Commission webpage: https://ec.europa.eu/eip/ageing/library/smart-age-friendly-cities-age-friendly-smartcities_en.

Wahl, H.-W., Iwarsson, S., & Oswald, F. (2012). Aging well and the environment: Toward an integrative model and research agenda for the future. *The Gerontologist, 52*(3), 306–316. https://doi.org/10.1093/geront/gnr154.

Wang, J., Chen, T., & Han, B. (2014). Does co-residence with adult children associate with better psychological well-being among the oldest old in China? *Aging & Mental Health, 18*(2), 232–239. https://doi.org/10.1080/13607863.2013.837143.

Washington State Department of Social and Health Services. (2019). *Common characteristics of people who neglect themselves*. https://www.dshs.wa.gov/altsa/home-and-community-services/self-neglect.

Waters, E., & Crandall, V. J. (1964). Social class and observed maternal behavior from 1940 to 1960. *Child Development*, *35*(4), 1021–1032. JSTOR. https://doi.org/10.2307/1126849.

Webber, S. C., Porter, M. M., & Menec, V. H. (2010). Mobility in older adults: A comprehensive framework. *The Gerontologist, 50*(4), 443–450. https://doi.org/10.1093/geront/gnq013.

World Health Organization (Ed.). (2007). *Global age-friendly cities: A guide*. World Health Organization.

Xinhua News Agency. (2018, November 30). *Experts say proposed measures to improve China's elderly care services*. http://www.xinhuanet.com/english/2018-11/30/c_137642185.htm.

Xinhua News Agency. (2019, March 21). *China continues to raise pensions for retirees*. http://www.xinhuanet.com/english/2019-03/21/c_137912554.htm.

Xu, H. (2019). Physical and mental health of Chinese grandparents caring for grandchildren and great-grandparents. *Social Science and Medicine, 229*, 106–116. https://doi.org/10.1016/j.socscimed.2018.05.047.

Ye, M., & Chen, Y. (2014). The influence of domestic living arrangement and neighborhood identity on mental health among urban Chinese elders. *Aging & Mental Health, 18*(1), 40–50. https://doi.org/10.1080/13607863.2013.837142.

Ye, M., Chen, Y., & Peng, Y. (2017). A new era in living arrangements: Determinants of quality of life among Chinese older adults. In M.-C. Tsai & W. Chen

(Eds.), *Family, work and wellbeing in Asia* (pp. 43–64). Springer Singapore. https://doi.org/10.1007/978-981-10-4313-0_3.

Ye, M., DeMaris, A., & Longmore, M. A. (2018). Role of marital quality in explaining depressive symptoms after marital termination among older adults. *Marriage & Family Review, 54*(1), 34–49. https://doi.org/10.1080/01494929.2017.1284703.

Ying, Z., Ning, L. D., & Xin, L. (2015). Relationship between built environment, physical activity, adiposity, and health in adults aged 46–80 in Shanghai, China. *Journal of Physical Activity and Health, 12*(4), 569–578. https://doi.org/10.1123/jpah.2013-0126.

Zeng, Y., Chen, H., Wang, Z., & Land, K. C. (2015). Implications of changes in households and living arrangements for future home-based care needs and costs for disabled elders in china. *Journal of Aging and Health, 27*(3), 519–550. https://doi.org/10.1177/0898264314552690.

Zhang, C., Cubbin, C., & Ci, Q. (2019). Parenting stress and mother–child playful interaction: The role of emotional support. *Journal of Family Studies, 25*(2), 101–115. https://doi.org/10.1080/13229400.2016.1200113.

Zimmerman, B. J., & Schunk, D. H. (2003). Albert Bandura: The scholar and his contributions to educational psychology. *Educational psychology: A century of contributions* (pp. 431–457). Mahwah: Lawrence Erlbaum Associates Publishers.

CHAPTER 3

The Development of Community Eldercare in Shanghai

This chapter reviews the development of home- and community-based long-term care service (HCBS) policies in Shanghai to understand macro-level influences in the community care ecology of the aging population. First, we introduce evolving Chinese eldercare and the challenges it poses to individual older adults, communities, and society. Next, we review caregiving alternatives for Chinese older adults and the history of the HCBS policies in Shanghai during the Economic Reform, paying special attention to the development of the HCBS program in the Jing'an District—the focus of this book. Finally, we analyze these eldercare policy evolutions from the social-ecological perspective.

3.1 Evolving Chinese Eldercare

Major changes in aging trends, urbanization, population mobility, the labor market, and family structure in China have challenged traditional family caregiving. We review how these macro-level factors influence the evolution of Chinese eldercare.

By the end of 2018, Shanghai's aging population (aged 60 and over) had reached 4.9 million, accounting for 33.6% of the total population, up from 4.5 million in 2016 (Shanghai Bureau of Statistics, 2019). There were 2281 older adults aged 100 and above in Shanghai in 2018, with 14.9 centenarians per 100,000 population in 2017 (Shanghai Observer, 2018a). Thus, the aging trend in Shanghai is both considerable and rapid.

L. Chen and M. Ye, *Community Eldercare Ecology in China*,
https://doi.org/10.1007/978-981-15-4960-1_3

Two simultaneous phenomena may relate to this rapid increase: declining fertility and increasing life expectancy. Because of the one-child policy in China, family size in Shanghai dropped from 2.8 persons per household in 2000 to 2.5 persons per household in 2014 (Sina, 2019). Over the same period, the natural population growth rate in Shanghai went from 3.51 in 1990 to −0.69 in 2017, steadily increasing the percentage of older adults in the population (Shanghai Bureau of Statistics, 2019). Moreover, the average life expectancy at birth in Shanghai has risen from 73.35 years in 1978 to 83.37 years in 2018: 85.85 years for women and 80.98 years for men (Shanghai Bureau of Statistics, 2019).

Shanghai's population is also aging at a much faster rate than in most developed countries. For example, the proportion of people over 65 years of age in the United States has doubled from 8% in 1950 to 16% in 2018 (Duffin, 2019). In Shanghai, the number of people aged 60 and over will account for 40% of the total registered population by 2030. By the end of 2050, it will reach 44.5%, exceeding Japan's 42.7%, and Shanghai will become one of the world's "oldest" cities (Yin, 2016). The current mandatory retirement age in China is 60 for men and 55 for women so that the working-age population will shrink (Liu & Sun, 2016).

At the same time, because of the one-child policy that started in the 1970s, many people are facing a "four-two-one" family structure now (Qi & Guo, 2007), i.e., two parents with a single child who also need to support four elderly parents while raising their child. According to Qi and Guo (2007), there will be 13.47 million such families by 2035. Because of the fertility policy and the mortality rate, the number of older adults who need to be taken care of by their family is increasing, while the number of family members available to take care of them is declining (Yin, 2016). China has one of the lowest child-dependency ratios (i.e., the ratio of people aged 0–14 to working-age people between 15 and 59) in the world but its old-age dependency ratio (i.e., the ratio of people aged 60 and over to working-age people) continues to rise. The total dependency ratio in Shanghai reached 58.8% in 2017, from 54.1% in 2016. In other words, among 15- to 59-year olds in the labor force, every 14.7 people out of 25 needed to take care of at least one older adult or one child in 2017 (Shanghai Observer, 2018a). Increasingly, an aging population is highly related to an increase in the burden of chronic diseases and long-term care (Word Health Organization & National Health and Family Planning Commission of China, 2016). By 2018, it was time for the Shanghai government to implement reliable and affordable measures to face this aging trend.

Urbanization has introduced significant migration from inner China to the east coast. The improvement of education, the development of technology, and the domestic immigration of the population, along with the progress of urbanization, has also changed the traditional eldercare system (Y. Zeng, Land, Gu, & Wang, 2014; J. Zhang, 2017). First, young people migrate to cities that are far away from their parents, because they have to take work outside of their home provinces to pursue ambitious careers. As a result, young people are no longer able to provide local support to their parents (J. Yang, 2012). Second, women's roles and personal expectations have changed. In 2010, the employment rate of women aged 18 to 64 was 71%, and 61% of them were in cities (Y. Wang, 2011). Employed women no longer take full responsibility for caring for their elderly parents (Summerfield, 1994). F. Chen and colleagues (2011) reported that older adults make a positive impact by taking care of their grandchildren (F. Chen, Liu, & Mair, 2011); however, the high cost of living and related healthcare costs make it difficult for adult children to live with their elderly parents and be fully responsible for their care (Ye, Chen, & Peng, 2017). As a result, the number of extended families (e.g., four generations living in the same household) has continuously decreased (Liu & Sun, 2016).

Meanwhile, the concept of providing family care for older adults has changed in China (L. Chen, 2016). Young people spend more time and energy on working and studying. They also focus on the quality of life, leisure, entertainment, and self-development, which means that they have limited time and energy to take care of their elderly parents. In addition, young people now value their personal space, no longer fully obey their parents' orders, and do not want to be fettered by the responsibility of caring for elderly parents for a long time (J. Zhang, 2017). Due to these social changes, family concepts and the status of older adults at home have been shaken (Liu & Sun, 2016). Because of the weakening of older adults' status at home after industrialization, it is difficult to maintain the traditional family-based eldercare system (Fei, 1939). China started its industrialization in the late nineteenth century when Western imperialist influence was high. Since then, the development of enterprises and the flow of population have dramatically changed the traditional family structure in China (Ma, 2005; M. Yang, 1994). As a consequence, older adults no longer hold the same level of authority at home, as their previous life experiences do not fit the industrialized world (Fei, 1939). Even when adult children are capable of caring for their elderly parents, many young people no longer see this as one of their responsibilities (Word Health Organization & National Health and Family Planning Commission of China, 2016). Furthermore, many elderly parents do not live with their adult children as previous generations did (Ye et al., 2017). These changes have directly affected the eldercare system and the economic security, quality of life, and mental health of older adults (L. Chen, Ye, & Kahana, 2019).

Traditionally, older adults in China were taken care of by their own families (L. Chen, 2016; Fei, 1939; Ye, 2008). The extended family served as a safety net within which the basic needs of all family members—including orphans, widows, the disabled, older adults, and the temporarily unemployed—could be met (Fei, 1939). In a typical extended family, three generations of members (grandparents, parents, and grandchildren) lived together and shared the family's property and income (Fei, 1939). This traditional family system placed great emphasis on the social role of older adults, and children had the greatest responsibility for taking care of their parents, although most of the care was expected to be done by the daughter-in-law since it was a patriarchal family structure (Chappell & Kusch, 2007; Yılmaz & Zeng, 2016; Zhan & Montgomery, 2003). Similar family structures can also be observed in South Korea, where traditional Chinese culture has had a great influence (Kim & Lee, 2003).

However, family size and structure in China have undergone tremendous changes after the Economic Reform and opening up of China (Ye et al., 2017). Before the 1950s, the average household size in China was 5.3. In 1990, it was reduced to 3.96, in 2010 to 3.10, and in 2012 to 3.02. By 2014, the family size in Shanghai was 2.5 (Sina, 2019). The proportion of extended families in the total number of households in China is also declining, from 48% in the 1990s to 13% in 2014 (Word Health Organization & National Health and Family Planning Commission of China, 2016). At the same time, the proportion of two-generation families (parents and children) is increasing. In the 1990s, about 48% of families were two-generation families, and only 5% were one-generation families. In 2010, one survey indicated that 80% of Chinese families were two- or one-generation families (Word Health Organization & National Health and Family Planning Commission of China, 2016; J. Yang, 2012).

A completely family-based eldercare system may lead to burdens for both caregivers and care receivers (E. Kahana & Young, 1990; Word

Health Organization & National Health and Family Planning Commission of China, 2016). One of the biggest challenges for older adults is their loss of physical function due to chronic diseases or age-related health conditions (Bury, 2000; E. Kahana & Kahana, 1996). One way to assess the physical function of older adults is to measure their independent daily activities (Deimling et al., 2017; Q. Feng et al., 2013). In 2010, 33 million people older than 60 (19% of this age group) reported difficulties in completing daily activities, and 11 million of them (6%) were fully dependent (L. Zhang, Curhan, Hu, Rimm, & Forman, 2011).

Older adults' quality of life is highly associated with their caregiving system (Z. Feng, Liu, Guan, & Mor, 2012). In other words, whom they can rely on to help with their daily activities is important for their quality of life (Sousa & Figueiredo, 2002). On the other hand, researchers have also found that the dependence of older adults on their caregivers may also cause care pressures for caregivers (B. Kahana, 2015; E. Kahana & Young, 1990). The heavy work of caregiving can cause many psychological problems for family caregivers, which in turn decreases the quality of life for older adults in their care (E. Kahana & Young, 1990). Providing care for older adults at home can also cause challenges for young people who work outside the home (J. Zhang, 2017). J. Yang (2012) found a significant negative correlation between the needs of older Chinese adults and the quality of life of their family caregivers.

3.2 Emerging Eldercare Alternatives in Shanghai

Healthy behaviors, family environment, and social factors account for 70% of healthy aging among older adults (World Health Organization & National Health and Family Planning Commission of China, 2016). Therefore, to improve the level of health among older adults and enhance the quality of life for both caregivers and care receivers, researchers should focus on the broader social factors and living environment to build an effective system (World Health Organization & National Health and Family Planning Commission of China, 2016). Currently, there are two alternatives to family care for older adults: institutional caregiving and community-based care. The following sections introduce the history and the challenges of institutional caregiving and community-based care policy in Shanghai.

Institutional caregiving refers to formal care available to older adults outside of the home. For example, as long as older adults can pay a certain monthly fee, they can live in nursing homes or long-term care facilities, which provide a place to live and care (Centers for Disease Control and Prevention, 2015). Because institutional care was advocated in the early 1990s, the Shanghai Municipal Government launched a policy advocating to establish a nursing home in each residential neighborhood to meet the evolving needs of eldercare (Ye, 2008). However, because of the large number of older adults, establishing a sufficient number of nursing homes and long-term care facilities was difficult to achieve (Ye, 2008). In 2018, there were 703 long-term care institutions in Shanghai, with a total of 144,000 beds, but this number could accommodate only 1.9% of the city's aging population (Yin, 2016).

Moreover, the cost of staying in a long-term care facility and the levels of care can vary widely. Costs range from 1600 *yuan* (i.e., 230 USD) per month to 30,000 *yuan* (i.e., 4300 USD) per month (Phoenix Weekly, 2017). In addition, market-oriented policies may further increase the cost of institutional care services and render many older adults unable to afford them (Wong & Leung, 2012). Therefore, market-led nursing home care models will only exacerbate the current inequities in access to and use of health services. Nursing homes in rural or suburban areas may cost less, but older adults are usually not satisfied with these facilities because of their distance from their original homes and some essential healthcare services, as most healthcare services are located in urban areas (Chu & Chi, 2008; Pan, Zhao, Wang, & Shi, 2016). Most ordinary working-class retirees face this dilemma; many of them choose to stay in their own homes, regardless of their increasing needs.

However, even if older adults can afford a nursing home that is close to their original home, they are generally not willing to move to a nursing home unless this decision is strongly supported by their adult children (L. Chen, 2016; L. Chen & Ye, 2013). Z. Li (2012) showed that many older adults think that being "served around the clock" by their children is the greatest blessing they can have. Moreover, to respect one's older parents is a virtue in Confucian culture. Because of this traditional Chinese view (i.e., filial piety), many hold the belief that only people who are abandoned by their children would enter nursing homes (Ye, 2008). For those who must leave their home to live in a nursing home or senior apartment housing, institutional life means losing everything they once knew and owned, creating a serious sense of loss. The authors' previous studies have shown that adult children have the same concerns when thinking of moving their parents to a nursing home (L. Chen, 2016). Adult children are afraid of the stigma of "not respecting filial piety" (L. Chen & Ye, 2013). For these reasons, the use rate of geriatric beds in Shanghai is not high. According to statistics from the Shanghai Municipal Statistics Bureau (2017), the use rate of geriatric beds was only 86% in 2017, accounting for less than 3% of the total aging population in Shanghai.

Because of increasingly unreliable family caregiving and the inadequacy of nursing home resources, community-based eldercare and relevant services become an alternative for older adults in Shanghai. In fact, recent studies have found that fewer older adults are willing to rely solely on their children for caregiving (L. Chen et al., 2019; Ye et al., 2017). Another recent study found that 80% of older adults in Shanghai chose to live alone or with their spouse only (Ye et al., 2017). According to T. Li (1997)'s survey, only 15.29% of middle-aged people (ages 40–55) in Shanghai were willing to live with their adult children after the children married. Another recent study found that 15.17% of older adults were willing to pay out of pocket to hire a nanny to take care of them (Ye et al., 2017). However, many older adults are unable to afford the cost of a nanny (Wong & Leung, 2012). Ye and colleagues (2017) also reported that 19.67% of their elderly respondents were willing to use publicly funded long-term care services, and about half wished to be cared for by their spouses (Ye et al., 2017). Because the limited resources of families and nursing homes cannot fully meet the needs of older adults for long-term care, community-based home care programs have developed (Jing'an News, 2015; Shanghai Observer, 2018b; F. Zhang, Ahrentzen, & Zhang, 2019).

3.3 The History of HCBS Policy in Shanghai

Before 1980, China had a planned economy (Tang & Parish, 2000), whose government-supported eldercare system targeted the mainly "three no" elderly (L. Chen & Ye, 2013), referring to those aged 60 and above who have no working ability, no resources, and no supporters, or their potential supporters themselves have no support capabilities (L. Chen, 2016). The eldercare system in the planned economy, therefore, aimed to provide social assistance to those in an impoverished situation (L. Chen, 2016).

The eldercare system was mainly provided by government-supported enterprises in Shanghai (Ye, 2008). Under the planned economy system, all organizations were related to government agencies, and enterprises and institutions were part of the entire administrative system (Xie, 2016). Because the state had almost all the resources, it would also cover social affairs, while organizations outside of the state had no resources to provide these services (Xie, 2016). The state, therefore, provided public services through administrative enterprises and institutions. When providing public services, such as the healthcare system and the old-age insurance system, these organizations and units were particularly aware of the role of the government in management (Xie, 2016). However, with the more recent Economic Reform, the government has pursued a socialist market economic system and started to take over these public services from the previous enterprises (Zhou, 2010).

In 1992, the State Council issued the Decision on Pension System Reform for Enterprise Employees (Ministry of Human Resources and Social Security of the People's Republic of China, 2012). In 1993, to reform the social security system, the state launched the Decision on Several Issues Regarding Building Up the Socialist Market Economic System to endorse a multichannel capital-source insurance system. In 1995, the Notice on Deepening Pension System Reform for Enterprise Employees ensured that each urban worker had a pension account. In 1997, the Decision on Building up a Unified Pension System for Enterprise Employee set up a rule that 11% of wages must be deposited into the employees' pension account (Zhou, 2010). This rule has become a guideline for the formal Chinese pension insurance system since then (Ye, 2008).

Prior to the Economic Reform, employees would not accumulate money in pension accounts. As a result, huge transition costs had to be paid by the government to make up for these gaps. In 2000, the National Council for Social Security Fund was established to address this gap in the national pension insurance through value-added operations (Ye, 2008). More recently, in 2019, the central government started to transfer 10% of the shares of state-owned enterprises to social insurance funds (People.cn, 2019).

Because the social security system that was implemented in 1993 is fragile and incomplete, Shanghai is currently facing both a social pension funding imbalance and a weak social welfare system. Therefore, elderly retired people cannot fully rely on pensions and welfare systems in their later life, since many of their needs cannot be met with the existing systems. As the family-based eldercare pattern, the nursing home, and the existing welfare system cannot fully cater to the needs of older adults, elderly services in the community have become an essential resource for them (Wu, Carter, Goins, & Cheng, 2005).

When the Economic Reform moved the country from a planned economy to a market economy, many public services in urban China started to be delivered through local communities (Wu et al., 2005). In Shanghai, each community has a residential committee. The residents organized the first residential committee in the early 1950s, when Shanghai residents worked in government-supported enterprises (Sangren, 1987; Summerfield, 1994). These enterprises served not only as places of business producing goods for society but also as social organizations for the employees (Ye, 2008). They replaced many of the functions of society and provided welfare services for employees (Summerfield, 1994). Once a person was hired, he or she would enter a universal social security system that covered almost all aspects of his or her life, such as health care, marriage, childcare, housing, insurance, and death. Sometimes, large enterprises would also offer schools for the children of its employees (Summerfield, 1994). Since almost all urban citizens were employees of such an enterprise, public services provided by communities targeted only those who were unable to work, such as people with disabilities. Other retired persons received support from their employers (Ye, 2008).

In the 1980s, in order to adjust to the reform of the urban economic system, the Ministry of Civil Affairs first advocated for community services, with the aim of providing welfare and convenience services for all community residents. During this period, community services undertook various social responsibilities that were increasingly released to society by *danwei*, as the previous system gradually disintegrated. At this time, the social welfare programs focused on those who are not employed by *danwei* (Summerfield, 1994).

With the gradual establishment of a highly centralized economic system, residential committees have become supporting organizations for residents who were not covered under the previous welfare system (Benewick, Tong, & Howell, 2004). The administrative expenses and personal allowances of the residential committees are borne uniformly by the subdistricts, and these allowances gradually became salaries. The administrative system from the planned economy period had insufficient team building and operational efficiency, and it was difficult to adapt it to public management and service functions under market conditions (Ye, 2008). The administrative level of the subdistrict is very low, and its financial resources, workforce, and capacity are limited. It naturally passes the pressure of its administrative work on to community organizations and the larger society (Xie, 2016; Ye, 2008).

Since the 1990s, the reform of urban communities in China has formed a new model. The Ministry of Civil Affairs of Shanghai asked each community to establish a community committee that is composed of the residents themselves. The committee is funded by the local government, which assists in delivering public services in the community (Ye, 2008). Community service has been listed as one of the pivotal projects of the Shanghai Municipal Government since then (Shanghai Municipal Construction Committee, 2016).

With the further deepening of the reform of Shanghai's economic system, the social responsibility of the community has increased. In 1995, the new system of "two-level government and three-level management" was implemented, and community services gradually formed a four-level network of cities, districts, subdistricts, and neighborhood committees (Sargeson, 2018). In 2006, the State Council issued the Opinions on Strengthening and Improving Community Service Work, which aimed to innovate service methods, broaden service areas, and strengthen service functions in the local communities. In 2007, the Shanghai Municipal Government issued the Implementation Opinions on Improving Community Services and Promoting Community Construction, which proposed improving community services and promoting community construction as an important basis for building a harmonious socialist society (Shanghai Municipal People's Government, 2007).

In 2000, the Shanghai Academy of Social Sciences conducted a survey on community services and asked, "According to your current situation, what do you think are the most needed services for older adults at the moment?" Residents responded, in order of priority: (a) medical services, (b) meal services, (c) grocery shopping services, and (d) fitness activities (C. Sun, 2000). The home- and community-based Service was created based on these priorities.

Since 2000, the Shanghai Municipal Bureau of Civil Affairs has piloted HCBS services in some subdistricts in Yangpu District and Pudong New District (Ye, 2008), providing both home- and community-based service (HCBS) centers and adult day care for seniors. These services have targeted older adults with disabilities, economically disadvantaged senior citizens, and retired workers who received awards from the government. The government set up a special fund to cover these expenditures through the welfare lottery (Ye, 2008).

Since the Shanghai Municipal government issued the regulation of home- and community-based services for older adults in 2003 (Ye, 2008; G. Zhang, 2015), public funds and support for community services have been allocated to cities, districts, and subdistricts at a ratio of 1:1:1. The funding provides subsidies for older adults who lack resources, retired workers who received awards from the government, those 80 years old and over, those who live alone, and centenarians. The subsidy standard is a maximum of 200 *yuan* per person per month (Ye, 2008). Also in 2003, the Shanghai Municipal Civil Affairs Bureau issued a policy to deepen the pilot work of HCBS services (Gui, Xu, Lou, & Tian, 2010). The policy clarified the management, organizational network, and service personnel of the home- and community-based service network (Ye, 2008) at three levels: governmental, intermediary organizational, and service entity. The governmental level refers to the responsibilities of the Social Welfare Division of the Civil Affairs Bureau, which is mainly responsible for formulating relevant policies, conducting macro-control, managing comprehensive coordination, and allocating funds. The intermediary level refers to an organization that is specifically commissioned by government departments to carry out home care services (Ye, 2008) at three levels of home care service centers in cities, districts, and subdistricts (Sargeson, 2018). Finally, the service level is entrusted by the government to implement service projects (Ye, 2008).

The Social Welfare Division formulates policies specifically around service entities in the community—organizations that specifically provide services to service targets, such as HCBS programs or associations of older adults. According to the new policy launched in 2003, the government regulates the workflow of HCBS and promotes the home care service vouchers to purchase services for economically disadvantaged older adults (Ye, 2008; Gui et al., 2010). The vouchers are issued to older adults through residential committees and are covered by public funds (Gui et al., 2010).

Since 2004, the Shanghai Municipal Government has for the first time included the home- and community-based project in the implementation project, and service funds have been included in the municipal budget (Ye, 2008). In addition to governmental funds, money has also been raised through society. Welfare lottery tickets in the urban area have totaled 10 million yuan, and city and district governments were expected to contribute funds—a 50/50 match—to demonstrate their support for the services (Ye, 2008). Moreover, the Shanghai Municipal Government had stipulated that rural surplus labor should be preferentially recruited as HCBS staff members (Zhou, 2010; G. Zhang, 2015) with a salary of 800 yuan per month, of which 200 yuan was for transportation costs (Ye, 2008). The HCBS program also provides social security for these staff members (Ye, 2008).

In 2004, the Civil Affairs Bureau set up a "90-7-3" framework for the eldercare plan (Shanghai Government, 2007), which was the backbone of the developing HCBS in Shanghai. The framework aimed to integrate healthcare services for older adults in three groups: 90% primarily relying on family caregiving; 7% seeking HCBS for care and assistance, in addition to family caregiving; and the remaining 3% possibly needing to move into nursing homes as their healthcare needs exceeded family caregiving capacities (Shanghai Government, 2007).

In 2005, the Municipal Government issued Document No. 42, expanding the scope of subsidies for home care services for older adults (Shanghai Civil Affairs Bureau, 2005; Ye, 2008). Those over the age of 60 were clearly defined as the target population for HCBS programs. The subsidy amount was 150–250 yuan per person per month: 150 yuan for the 60- to 79-year olds; 200 for the 80- to 89-year olds; and 250 yuan for those 90 years old and above (Ye, 2008). Home-based pension subsidies can also be brought into institutions, so older adults who live in residential care institutions can still receive home subsidies if they meet the maximum income requirements (Ye, 2008).

In 2006, the Municipal Government issued a new policy to regulate who could receive services purchased by the government (Shanghai Civil Affairs Bureau, 2006; State Council Information Office of the People's Republic of China, 2006). According to the new policy, Chinese older adults who lived overseas were no longer included in the subsidy. In addition, older adults who wanted to apply for services would receive an assessment from the government. The assessment was based on their health conditions and economic status, including self-care capability, cognitive capability, mental health status, visual ability, living environment, and diseases (State Council Information Office of the People's Republic of China, 2006). Each applicant received a score represented as normal, light, medium, or heavy, and the person received services or financial support accordingly. For example, an older adult with a score of heavy would receive 400 Yuan per month for home care services (State Council Information Office of the People's Republic of China, 2006).

The subdistrict governments of Shanghai provide HCBS programs (e.g., meal services or bathing services) that serve low-income older adults. Other older adults can also pay out of pocket for these services (Sina, 2019). Since 2004, a large number of older adults have had HCBS services purchased by the Shanghai government. In addition, the number of elderly clients has increased significantly. For example, the government purchased bathing services for 100,000 older adults in the winter of 2004 (Shanghai Yearbook Office, 2005). In the winter of 2005, 111,000 older adults had bathing services purchased by the government (Shanghai Yearbook Office, 2006). In the winter of 2006, the number increased to 180,000 (Shanghai Yearbook Office, 2007). It is expected that in the future, the government of Shanghai will continue to increase spending on HCBS services.

3.4 The HCBS Program in the Old Jing'an District

By the end of 2007, the total number of older adults in the Jing'an District had exceeded 74,000, accounting for 24.04% of the total population of the district (Jing'an Government, 2007). In order to meet the needs of one-fourth of the population in Jing'an District, the district government decided to design an eldercare system based on community- and home-based eldercare programs, with institutional care as a supplement. Directed by the Jing'an District Party Committee and the District Government, the construction of an HCBS program for older adults became the district's highest priority in 2007 (Jing'an Government, 2007).

Because the Jing'an District is located in the city center and is a high-end commercial service center, real estate prices in the area are very high (Anjuke, 2019). The district government has a limited budget for building large nursing homes. Also, given its large aging population, it was wise for the Jing'an District government to seek community-based programs as a viable alternative for eldercare.

Before community-based services, home care services for community-dwelling older adults were basically from the nanny market to provide home-based caregivers (i.e., *baomu*, Wu et al., 2005). In other words, the older adults used nanny from a nanny agency to come to serve them. However, because of the low wage threshold for nanny work, the quality of the nanny staff varies. The nanny agency was a for-profit organization that needed to make money. Since older adults had limited budgets to purchase services, the quality of services provided by the nanny was generally low. Because the nanny market was unregulated at the time, in 2006, the old-old (i.e., aged 80 and over) and older adults who had difficulty living on their own could receive some form of home care services from the district government, outsourced through a nonprofit, community-based organization.

In 2006, the Department of Aging in the Jing'an District conducted a survey of elderly clients to evaluate the quality of care purchased by the government from the nanny market and clients' service satisfaction, including questions such as "How do you feel about the service?" and "Would you recommend the service to others?" The clients reported the quality of service was not high and that the staff members from these vouchers were unprofessional. For example, clients pointed out that sometimes the staff did not come on time. Because the regulation was not strict, some staff members worked less time but asked for a full salary.

Another problem was that since the contracts with the staff members were not clear, their responsibilities were ambiguous. The discrepancies between older adults' and staff members' expectations created challenges for service delivery and quality of care. Because of these problems, the Jing'an District government decided to establish its own community-based eldercare program.

As a result, the original HCBS program was affiliated with the Department of Aging of the Jing'an District. In 2005, the government allocated 200,000 *yuan*, set up a working group, recruited staff, and organized a service team. The district funding for this activity was allocated to the subdistrict governments, and the Department of Aging managers each subdistrict recruited staff to provide services. In doing so, the Jing'an District government expected to attract more investment in community-based eldercare from third-sector and nonprofit organizations. For example, some HCBS service centers have a service support group that links services in the centers to nearby enterprises and businesses—e.g., hospitals, clinics, barbers, laundry shops, real estate companies, and restaurants. These businesses send their employees to the service centers to provide cleaning, bathing, meal, medical, and emergency services for elderly clients. All of these services are usually negotiated below the market price, making them affordable for elderly clients.

Every year, the Jing'an District government not only subsidizes 100,000 yuan per year for each service center for daily maintenance, but also subsidizes 50,000 yuan per service center for operations (e.g., staff salaries) each year. The five subdistrict governments also subsidize another 50,000 yuan per service center per year. This financial support ensures the daily operation of each service center and low service prices, making them accessible to all elderly clients.

In 2008, a year after its emergence, the Department of Aging in the Jing'an District conducted a survey to examine elderly clients' reception of the HCBS program. Both well-off and disadvantaged older adults enjoyed the program's services; in particular, they highly praised the low cost, which was designed to be cheaper at the centers than on the general market. Because of the small profits and large number of clients, this business model was widely adopted throughout the HCBS program.

Since 2009, the Jing'an District government has no longer directly administered the HCBS program but has allowed its five subdistrict governments to operate autonomously and develop service centers. The district government monitors and evaluates the quality of care of each service center annually.

After one year of operation, the HCBS program determined that the greatest need among elderly clients was the meal service. Because most elderly clients wanted to start lunch at around 11 to noon, with limited service staff members, there was a significant supply and demand contradiction. Therefore, the HCBS agency set up a site for serving meals in a location near the community so that resources could be concentrated and services (e.g., preparing meals) offered at this site together. Since 2007, the whole district has promoted the HCBS service centers. With only two in 2007, the number of centers reached 37 in the Jing'an District in 2015. All older adults aged 60 and over, living in the Jing'an District, can enroll in the HCBS program through the Civil Affairs office in each subdistrict. Most services are free; however, some, such as the meal service and the assistance service, charge low fees (e.g., each meal costs about \$1.50). Older adults who have extremely low income, live alone, and/or have disabilities receive coupons from the district government to enroll in the program for free.

The HCBS program offers 10 types of services at all the service centers:

- Assistance service: Staff helps elderly clients to run errands, including making purchases, picking up packages, or paying for utilities;
- Bathing assistance: Each service center installs shower stalls, and staff members help elderly clients to bathe;
- Companion service: Staff members accompany elderly clients for some outdoor activities;
- Consulting and counseling service: Professional psychotherapists, lawyers, or social workers are invited to service centers to provide psychological counseling and legal consulting for older adults in the neighborhood, regardless if they are elderly clients;
- Meal service: Each service center provides lunch and/or dinner for enrolled elderly clients, with flexible meal plans. Elderly clients can order the number of meals at the beginning of every month or pay per meal as they go. They can also choose to have meals at the service center, take-home, or free delivery to the home;
- Emergency service: Staff members help elderly clients to deal with emergencies, such as fall and medical emergencies;
- Entertainment service: Staff members organize community activities, festivals, art exhibitions, or sports for elderly clients;
- Education service: Staff members hold lectures (e.g., health promotion) and courses (e.g., calligraphy, computer), or reading clubs for elderly clients;
- Medical service: Staff members invite professional physicians to provide medical consulting for elderly clients;
- Personal care service: Staff members outsource hairdressers and manicurists for elderly clients in the service center.

These service items aim to maintain elderly clients' independent living, both physically and financially, and to serve as many older adults in the community as possible. So in terms of allocating the service centers, the HCBS program conducted a study and found out that the radius of older adults' activities cannot be too large. If there are few service centers, it will be inconvenient for the older adults to visit the service centers because they would need to walk a long way. Therefore, the design is that a density of about two to three neighborhood committees has one service center, which is about ten minutes' walk for ordinary people. This distribution allows the service network to achieve full coverage. Each service center has two frontline staff members and one manager, as well as a management committee, whose members are neighborhood and elderly client representatives.

The HCBS program recruited those who are "4050" priority as the frontline staff for the service centers (G. Zhang, 2015). The "4050" group refers to laid-off male workers who were over 50 years old and female workers over 40 years (Yan, 2006). The staff members of the service centers are all local laid-off employees with Shanghai household registration. After such employees are trained, a government-certified employment permit is used to provide services. All basic staff wages are paid by the Finance Bureau, which grants subsidies on an hourly basis. The salary of the staff is the minimum wage in Shanghai. For example, at the beginning of its establishment in 2007, the frontline service staff of HCBS received 800 yuan per month, including 200 yuan for transportation and meal expenses, and 600 yuan for salary. The service center also pays social insurance for these workers.

This HCBS program in the Jing'an District relies on a third-party evaluation system to monitor its quality of care. Because the HCBS project was a new project for the government and the older adults in Jing'an District at the time, whether the program reaches its goal of helping older adults to receive proper services was a question.

The Dept. of Aging in Jing'an District formulated the evaluation measures for the evaluation of the HCBS service centers in Jing'an District. The Gerontological Society of Shanghai (GSS) has conducted surveys to evaluate the HCBS service centers in Jing'an District since 2008. The Secretariat of the GSS has set up a review department, drawing on the "Hong Kong Home for older adults Home Assessment Scheme" by the Hong Kong Senior Citizens Association, and has formulated an evaluation implementation plan. The Evaluation Department recruited six senior technology workers with management experience and four senior community workers as volunteers to conduct the evaluation. After one and a half years of the HCBS program began, the evaluation began in June 2009. In order to ensure the authenticity and objectivity of the evaluation, and to reflect the fairness and openness of the evaluation, the HCBS program set up a suggestion box in each service center, a website link, and a dedicated email address. This evaluation system enables interaction and communication between HCBS directors, service center staff, and elderly clients (Gerontological Society of Shanghai, 2009a).

The evaluation plan is divided into field service and user satisfaction. The field service evaluation is conducted in the form of "mystery visitors," and the monthly field evaluation is performed, and the evaluation report is submitted. Field items include evaluations of six major categories with 100 items, such as "Is there a non-slip mat at the bathroom floor?" The mystery guest goes to various sites to observe the quality of service facilities and the quality of services the staff offers (Gerontological Society of Shanghai, 2009a).

User satisfaction is evaluated once a year. Every October, the Evaluation Department of the Secretariat of the GSS forms a Satisfaction Evaluation task group to formulate the implementation plan, design the Annual Satisfaction Assessment questionnaire, conduct surveys, and report to the department to Jing'an District. The research team mails questionnaires, conducts telephone interviews, and also holds focus groups to allow older adults to express their feelings, opinions, and suggestions. Questions include "Would you recommend the program to other older adults?" and "Overall, are you satisfied with the service you received from the HCBS service center?" The review team submits an Annual Satisfaction Report to the Jing'an District Department of Aging (Gerontological Society of Shanghai, 2009a).

The evaluation report mainly reflects the quality of the service center and elderly clients' views on the service center, including the quality of the services and the staff members' attitudes. According to the reports over the years, elderly clients are generally satisfied with the services provided by the HCBS service centers and are willing to recommend the program to future users (Ye & Chen, 2014). The meal service has received relatively lower satisfaction ratings among the ten services because older adults had high expectations and different tastes in food; however, the meal service is the one that the clients purchased the most (Gerontological Society of Shanghai, 2009b).

Elderly clients reported that they expected to use more bathing assistance and emergency services in the future, suggesting their increasing needs for ADL services (Gerontological Society of Shanghai, 2009b). Similarly, older adults' experiences of utilizing ADL services, such as meal service and bathing assistance, are more likely to determine the overall program satisfaction than other services (Ye, 2010). Elderly clients can also file complaints, suggestions, and general impressions through the annual evaluation. The HCBS program adjusts its services based on this feedback. For example, in 2013, several service centers began to provide breakfast based on elderly clients' requests. Since 2017, most of the centers have implemented year-round service, including the Lunar New Year holiday, extending the existing previous service schedule.

The evaluation survey represents the beginning of consumer-directed approaches in Shanghai's community-based eldercare, allowing elderly clients to set the standards of quality for services (Benjamin & Fennell, 2007). The strategy of adjusting services according to evaluation surveys is particularly meaningful for healthcare services. Since older adults know best about the services, such as home care and personal care, consumer-directed approaches can achieve greater satisfaction and better outcomes than those of traditional approaches based on standards set by professionals (Benjamin & Fennell, 2007).

3.5 Understanding Macro-Level Influences for HCBS in Shanghai

Based on the history of Shanghai community eldercare policy and the HCBS program in the Jing'an District, changes in the macro environment, such as the implementation of policies and evolving sociocultural contexts, have a direct impact on the development of HCBS in Shanghai.

Policies related to eldercare in Shanghai and Jing'an District were enacted in the context of the decline of the traditional home care culture. Considering filial piety as the core value of Confucianism in Chinese culture and its strong influence, many Chinese older adults will not choose to move into a nursing home or at least do not consider moving into a nursing home the best option for the older adults (L. Chen & Ye, 2013). When older adults think that the next generation cannot provide care for them, or when they do not want to be a burden on their adult children, about 60% of them will seek institutional care (Wong & Leung, 2012). In general, caring for older adults has been considered the most important responsibility of the family (Y. Sun, 2017). However, the sustainability of this form of care is changing as demographics change and is becoming less common (Z. Feng et al., 2012). In addition, because of changes in population structure and social transformation, filial piety has weakened as the core value of Chinese culture. In modern Chinese society, people show their filial piety in different ways, and caring for older adults is only one of them. For example, Y. Sun (2017) conducted interviews among 24 Beijing adults with an average age of 21 and found out that acceptance of different lifestyles between elderly parents and adult children is a way of showing filial piety. At the same time, the availability of family care for older adults also varies depending on the family situation (L. Chen, 2016).

Constructing and maintaining a new concept of filial piety in Chinese society and transforming healthcare services require government investment, interagency cooperation, and the establishment of a government-supervised long-term care system (Ikels & Stanford, 2004). Global evidence suggests that home caregivers need adequate support to fulfill their responsibilities (Word Health Organization & National Health and Family Planning Commission of China, 2016). When a full array of care options is available, home caregivers can fully meet their care responsibilities (Rickerson et al., 2005), and family caregivers still play a key role in China. In the absence of a sound long-term care service system, family caregivers are expected to bear a heavier psychological, emotional, and physical burden when they assume responsibility for the care of their family members (Rickerson et al., 2005). If home caregivers are unable to properly handle complex care activities, the quality of care will be greatly reduced, and the dignity of the care recipients will suffer (Black et al., 2013; World Health Organization, 2016). Furthermore, the lack of care services in the public system will have a negative impact on family relationships and ultimately social harmony (Huang & Fahmy, 2013). Therefore, measures at the community level are needed to strengthen family care for older adults and to promote solidarity across society to support these measures and reshape filial piety.

Policies related to HCBS in Shanghai and Jing'an District were also associated with the rise of the market economy in Shanghai. Looking back, we can see that the provision of eldercare services in Shanghai has undergone a process of transition from a planned economy to a market economy. In this transformation, we can observe that the rise of the market economy has given social forces more room and resources to participate in the provision of eldercare services. The logic of the market economy and the diversified demands it brings play a role in the payment methods and content of services. On the other hand, because of the non-excludability and low profit of public services, social public services, such as HCBS programs, need the support of government resources (Wallis & Dollery, 2001).

Since the aging population in Shanghai continues to grow, the Shanghai government faces the financial challenge of providing all services for its elderly residents (Yin, 2019). It is impossible to rely entirely on the market to provide public goods (e.g., eldercare services; Wallis & Dollery, 2001). For example, activity/recreation facilities for community-dwelling older adults are free and open to all older adults living in the community. Then older adults tend not to spend extra money to purchase activity equipment themselves. This behavior is called free riding, which suggests public services and facilities may not be profitable through the market (Winston, 2006). Whereas, the government emphasizes representativeness and fairness in the provision of public services, it cannot meet diverse needs among individuals (Wallis & Dollery, 2001), which adds on extra pressure to policymakers and public service providers (Winston, 2006).

Community-based organizations can help the government to balance its financial and service burden (Boyer, Slyke, & Rogers, 2015; Hodge & Greve, 2010). The emerging policy tools of dismantling have become one of the wonderful remedies to solve the "welfare state crisis" that is common in developed countries. For example, the HCBS program targets older adults in the local communities, in order to flexibly adjust their service strategies based on elderly clients' needs and requests.

The decentralization of eldercare provision is a critical step for the Shanghai municipal government to avoid having a "welfare state crisis" in the future. The Shanghai municipal government requires each district government to design its own HCBS programs and relevant policies to meet specific local needs. The municipal government and the Civil Affairs Bureau evaluate the HCBS performance in each district by measuring the number of elderly clients and the number of service centers. The municipal government offers incentives to encourage the development of tailored HCBS in each district and to expand service coverage (X. Zeng, 2010). The district government outsources service provision to community organizations who know their older adults better. The HCBS program in the Jing'an District has been a product of government purchasing. This collaboration between government and community has become a model for HCBS development in Shanghai (Shanghai Civil Affairs Bureau, 2006).

Another macro-level influence on the HCBS in Shanghai was the "4050" phenomenon. Many of the staff members were previously laid-off workers from state-owned enterprises. Because the Chinese government started the economic reform from a planned market to an economic market in the 1990s, many state-owned enterprises were either restructured or closed (B. Wang & Vongalis-Macrow, 2012; Yan, 2006; Zhao, 2002). One of the results of this reform was having around 17.24

million laid-off workers in 1998 (Gomersall & Wang, 2013). These laid-off workers were called "40-50," because when they were laid-off, they were almost forty or fifty years old (B. Wang & Vongalis-Macrow, 2012; Yan, 2006; Zhao, 2002). These workers provided a rich resource in the labor market, such as recruiting a staff member in an HCBS program (G. Zhang, 2015). In fact, the Shanghai municipal government launched a policy in 2004 to promote employment through HCBS and communities (Shanghai Government, 2005). The policy targeted the laid-off workers to get reemployed by providing sufficient training and opening community-based programs which can hire these trained workers (Shanghai Government, 2005).

Such a large-scale layoff, however, is a special historical event that exists only for a period of time. In fact, working people who are forty or fifty years old now do not see themselves got laid-off by any government policy in the future. The participants pointed out that they did not think the next generation with their similar age would join their team to become staff members in the HCBS program.

3.6 Chapter Summary

This chapter reviews the history of community-based eldercare policy in Shanghai and the development of the HCBS program in the Jing'an District, illuminating the macro-level influences on the community care ecology. Because of the growing number of older adults, Shanghai faces a severe burden of caregiving. Since it is difficult to rely on traditional family care alone, the Shanghai Municipal Government has attempted different policies focusing on nursing home care and community-based eldercare as alternatives to family caregiving. A series of HCBS eldercare policies has been in effect to supplement declining family caregiving and facilitate aging in communities. The Shanghai Municipal Government has implemented an inclusive community-based care policy, represented by the HCBS program in the Jing'an District. The HCBS program tries to offer as many services as possible to facilitate community-dwelling older adults' independence, including meal service, personal care, recreation, counseling, and legal aid. Moreover, the program has a third-party evaluation system to help monitor its quality of care and improve its services. Societal, economic, cultural, and demographic evolutions have brought multilayered influences on the macro level of the community care ecology.

References

Anjuke. (2019). *Jingan 2019 real estate price*. https://shanghai.anjuke.com/market/jingan/.

Benewick, R., Tong, I., & Howell, J. (2004). Self-governance and community: A preliminary comparison between villagers' committees and urban community councils. *China Information, 18*(1), 11–28. https://doi.org/10.1177/0920203X04040361.

Benjamin, A., & Fennell, M. (2007). Putting the consumer first: An introduction and overview. *Health Services Research, 42,* 353–361.

Black, B. S., Johnston, D., Rabins, P. V., Morrison, A., Lyketsos, C., & Samus, Q. M. (2013). Unmet needs of community-residing persons with dementia and their informal caregivers: Findings from the maximizing independence at home study. *Journal of the American Geriatrics Society, 61*(12), 2087–2095. https://doi.org/10.1111/jgs.12549.

Boyer, E. J., Slyke, D. M. V., & Rogers, J. D. (2015). An empirical examination of public involvement in public-private partnerships: Qualifying the benefits of public involvement in PPPS. *Journal of Public Administration Research and Theory*. https://doi.org/10.1093/jopart/muv008.

Bury, M. (2000). On chronic illness and disability. In C. E. Bird, P. Conrad, & A. M. Fremont (Eds.), *Handbook of medical sociology* (5th ed., pp. 173–183). Upper Saddle River: Prentice Hall.

Centers for Disease Control and Prevention. (2015). *Nursing home care*. http://www.cdc.gov/nchs/fastats/nursing-home-care.htm.

Chappell, N. L., & Kusch, K. (2007). The gendered nature of filial piety—A study among Chinese Canadians. *Journal of Cross-Cultural Gerontology, 22*(1), 29–45. https://doi.org/10.1007/s10823-006-9011-5.

Chen, F., Liu, G., & Mair, C. A. (2011). Intergenerational ties in context: Grandparents caring for grandchildren in China. *Social Forces, 90*(2), 571–594.

Chen, L. (2016). Children parenting: First and last adventure. In *Evolving eldercare in contemporary China* (pp. 95–107). Palgrave Macmillan US. https://doi.org/10.1057/978-1-137-54440-7_6.

Chen, L., & Ye, M. (2013). The role of children's support in elders' decisions to live in a yanglaoyuan (residential long-term care): Children's support and decision to live in a yanglaoyuan. *Journal of Cross-Cultural Gerontology, 28*(1), 75–87. https://doi.org/10.1007/s10823-012-9185-y.

Chen, L., Ye, M., & Kahana, E. (2019). A self-reliant umbrella: Defining successful aging among the old-old (80+) in Shanghai. *Journal of Applied Gerontology*. https://doi.org/10.1177/0733464819842500.

Chu, L.-W., & Chi, I. (2008). Nursing homes in China. *Journal of the American Medical Directors Association, 9*(4), 237–243. https://doi.org/10.1016/j.jamda.2008.01.008.

Deimling, G. T., Pappada, H., Ye, M., Nalepa, E., Ciaralli, S., Phelps, E., & Burant, C. J. (2017). Factors affecting perceptions of disability and self-rated health among older adult, long-term cancer survivors. *Journal of Aging and Health*. https://doi.org/10.1177/0898264317745745.

Duffin, E. (2019, October 17). *U.S.—Seniors as a percentage of the population 2050*. Statista. https://www.statista.com/statistics/457822/share-of-old-age-population-in-the-total-us-population/.

Fei, H. T. (1939). *Peasant life in China*. London: Hesperides Press.

Feng, Q., Zhen, Z., Gu, D., Wu, B., Duncan, P. W., & Purser, J. L. (2013). Trends in ADL and IADL disability in community-dwelling older adults in Shanghai, China, 1998–2008. *The Journals of Gerontology Series B: Psychological Sciences and Social Sciences, 68*(3), 476–485. https://doi.org/10.1093/geronb/gbt012.

Feng, Z., Liu, C., Guan, X., & Mor, V. (2012). China's rapidly aging population creates policy challenges in shaping a viable long-term care system. *Health Affairs,* 2764–2773. https://doi.org/10.1377/hlthaff.2012.0535.

Gerontological Society of Shanghai. (2009a). *New community-based long-term care program for older adults*. http://www.shanghaigss.org.cn/news_view.asp?newsid=6440.

Gerontological Society of Shanghai. (2009b). *Working report: Make the gerontological society of Shanghai as a third party*. http://www.shanghaigss.org.cn/news_view.asp?newsid=6453.

Gomersall, K., & Wang, M. (2013). Life-course experiences of laid-off workers and their attitudes towards migrant workers in urban China. *International Development Planning Review, 35*(3), 219–240. https://doi.org/10.3828/idpr.2013.15.

Gui, S., Xu, Y., Lou, W., & Tian, Q. (2010). Building elderly-friendly communities: An empirical study from shanghai. *Population Journal, 4,* 23–30. https://doi.org/10.3969/j.issn.1004-129X.2010.04.004.

Hodge, G., & Greve, C. (2010). Public-private partnerships: Governance scheme or language game? *Australian Journal of Public Administration, 69,* S8–S22. https://doi.org/10.1111/j.1467-8500.2009.00659.x.

Huang, Y., & Fahmy, S. (2013). Picturing a journey of protest or a journey of harmony? Comparing the visual framing of the 2008 Olympic torch relay in the US versus the Chinese press. *Media, War & Conflict, 6*(3), 191–206. https://doi.org/10.1177/1750635213495601.

Ikels, C., & Stanford, C. (2004). *Filial piety: Practice and discourse in contemporary East Asia*. Stanford: Stanford University Press.

Jing'an Government. (2007). *2006–2007 Jing'an district economic and social report*. http://www.jingan.gov.cn/jaxxgk/qfgw/jjgk/200704/t20070421_18002.htm.

Jing'an News. (2015, September 29). *Community-based eldercare services in summer.* Jing'an News. http://shuzibao.jingan.gov.cn/html/2015-09/29/content_327258.htm.

Kahana, B. (2015). Stress. In G. A. Colditz (Ed.), *The Sage encyclopedia of cancer and society* (pp. 1110–1112). Sage. http://knowledge.sagepub.com/view/the-sage-encyclopedia-of-cancer-and-society/i14133.xml.

Kahana, E., & Kahana, B. (1996). Conceptual and empirical advances in understanding aging well through proactive adaptation. In *Adulthood and aging: Research on continuities and discontinuities* (pp. 18–40). New York: Springer.

Kahana, E., & Young, R. (1990). Clarifying the caregiving paradigm: Challenges for the future. In D. E. Blegel & A. Blum (Eds.), *Aging and caregiving: Theory, research and practice* (pp. 76–97). Newbury Park: Sage.

Kim, J.-S., & Lee, E.-H. (2003). Cultural and noncultural predictors of health outcomes in Korean daughter and daughter-in-law caregivers. *Public Health Nursing, 20*(2), 111–119. https://doi.org/10.1046/j.1525-1446.2003.20205.x.

Li, T. (1997, October 21). *A preliminary study on supporting aging-in-place.* International Symposium on Aging in the 21st Century, Shanghai, China.

Li, Z. (2012). *The basic old-age insurance of China: Challenges and countermeasures.* World Pension Summit. http://www.worldpensionsummit.com/Portals/6/Zhen%20Li_Basic%20old%20age%20insurance%20in%20China.pdf.

Liu, T., & Sun, L. (2016). Pension reform in China. *Journal of Aging & Social Policy, 28*(1), 15–28. https://doi.org/10.1080/08959420.2016.1111725.

Ma, D. (2005, September). *Shanghai-based industrialization in the early 20th century: A quantitative and institutional analysis.* Monograph, Istanbul, Turkey. Retrieved from http://www.lse.ac.uk/collections/economicHistory/.

Ministry of Human Resources and Social Security of the People's Republic of China. (2012). *Explanation of social insurance legislation of the People's Republic of China.* Retrieved from http://www.mohrss.gov.cn/SYrlzyhshbzb/rdzt/syshehuibaoxianfa/bxffaguijijiedu/201208/t20120806_28572.htm [Chinese].

Pan, J., Zhao, H., Wang, X., & Shi, X. (2016). Assessing spatial access to public and private hospitals in Sichuan, China: The influence of the private sector on the healthcare geography in China. *Social Science and Medicine, 170,* 35–45. https://doi.org/10.1016/j.socscimed.2016.09.042.

People.cn. (2019, November 6). *Jing'an District: Developing older adults food support service.* http://sh.people.com.cn/BIG5/n2/2019/1106/c134768-33512275.html.

Phoenix Weekly. (2017). *Price for nursing home for elderly parents of middle class.* http://inews.ifeng.com/51518108/news.shtml?&back.

Qi, X. F., & Guo, Z. W. (2007). Micro simulation models for "4-2-1" families and their applications. *Population Research, 31*(3), 32–40.

Rickerson, E. M., Somers, C., Allen, C. M., Lewis, B., Strumpf, N., & Casarett, D. J. (2005). How well are we caring for caregivers? Prevalence of grief-related symptoms and need for bereavement support among long-term care staff. *Journal of Pain and Symptom Management, 30*(3), 227–233. https://doi.org/10.1016/j.jpainsymman.2005.04.005.

Sangren, P. (1987). *History and magical power in a Chinese community*. Stanford: Stanford University Press.

Sargeson, S. (2018). Grounds for self-government? Changes in land ownership and democratic participation in Chinese communities. *The Journal of Peasant Studies, 45*(2), 321–346. https://doi.org/10.1080/03066150.2016.1197909.

Shanghai Bureau of Statistics. (2019). *Shanghai statistical yearbook 2018*. http://tjj.sh.gov.cn/html/sjfb/201901/1003014.html.

Shanghai Civil Affairs Bureau. (2005). *Accelerating the development of the elderly care service in the "11th Five-Year Plan" period in Pudong New Area*. http://old.pudong.gov.cn/website/html/pdzw/newgovOpen_LDBZJA_ghjh/Info/Detail_200510.htm.

Shanghai Civil Affairs Bureau. (2006). *Promoting the development of the elderly care services in Shanghai*. http://www.shweilao.cn/cms/cmsDetail?uuid=05a1adbf-bbdc-479e-9c78-f78156fd80e3.

Shanghai Government. (2005). *Shanghai municipal government work report 2004*. http://district.ce.cn/newarea/roll/201203/26/t20120326_23187894.shtml.

Shanghai Government. (2007). *Shanghai government dedicates to "90-7-3" elder-care framework*. Retrieved from http://www.shanghai.gov.cn/shanghai/node2314/node2315/node4411/userobject21ai238364.html [Chinese].

Shanghai Municipal Construction Committee. (2016). *Adding elevators for multi-stories residential buildings in Shanghai*. http://www.shqp.gov.cn/housing/zhl/20190505/495118.html.

Shanghai Municipal People's Government. (2007, February 7). *Shanghai Municipal People's Government on the 'Eleventh Five-Year Plan' for the development of Shanghai's Civil Administration*. Retrieved from http://www.shanghai.gov.cn/nw2/nw2314/nw2319/nw10800/nw11407/nw16795/u26aw9814.html [Chinese].

Shanghai Municipal Statistics Bureau. (2017). *Shanghai yellow book 2016—Population—Age by district*. http://www.stats-sh.gov.cn/tjnj/nj16.htm?d1=2016tjnj/C0206.htm.

Shanghai Observer. (2018a, October 16). *There are currently 2281 older adults aged 100 and above in Shanghai. Last year, there were 14.9 centenarians per 100,000 population*. http://www.shanghai.gov.cn/nw2/nw2314/nw2315/nw4411/u21aw1344627.html.

Shanghai Observer. (2018b, December 19). *Those who bathed in the "Old Town" in Shanghai*. https://www.shobserver.com/news/detail?id=122558.

Shanghai Yearbook Office. (2005, April 28). *Shanghai yearbook 2004*. Shanghai Local History Office. http://www.shtong.gov.cn/dfz_web/DFZ/Info?idnode=72880&tableName=userobject1a&id=84184.

Shanghai Yearbook Office. (2006, June 25). *Shanghai yearbook 2005*. Shanghai Local History Office. http://www.shtong.gov.cn/dfz_web/DFZ/Info?idnode=76650&tableName=userobject1a&id=93693.

Shanghai Yearbook Office. (2007, October 4). *Shanghai yearbook 2006*. Shanghai Local History Office. http://www.shtong.gov.cn/dfz_web/DFZ/Info?idnode=79023&tableName=userobject1a&id=99879Sina. (2014). *Family size in Shanghai shrinks to 2.5 on average 0.3 fewer than in 2000*. http://sh.sina.com.cn/news/b/2014-05-15/082294878.html?from=sh_cnxh.

Sina. (2019). *Guidelines for applying for HCBS services and subsidies*. Sina.Com. http://sh.sina.com.cn/life/ylzn/2019-04-08/detail-ihsxncvh7915989.shtml.

Sousa, L., & Figueiredo, D. (2002). Dependence and independence among old persons—Realities and myths. *Reviews in Clinical Gerontology, 12*(3), 269–273. https://doi.org/10.1017/S095925980201239X.

State Council Information Office of the People's Republic of China. (2006, November 29). *Shanghai government report on elderly care services*. http://www.scio.gov.cn/XWFBH/gssxwfbh/xwfbh/shanghai/Document/320847/320847.htm.

Summerfield, G. (1994). Economic reform and the employment of Chinese women. *Journal of Economic Issues, 28*(3), 715–732. https://doi.org/10.1080/00213624.1994.11505579.

Sun, C. (2000). A study on the leisure life of urban older adults: A case study in shanghai. *Academic Quarterly of Shanghai Academy of Social Sciences, 3*, 126–134.

Sun, Y. (2017). Among a hundred good virtues, filial piety is the first: Contemporary moral discourses on filial piety in urban China. *Anthropological Quarterly, 90*(3), 771–799. https://doi.org/10.1353/anq.2017.0043.

Tang, W., & Parish, W. L. (2000). *Chinese urban life under reform: The changing social contract*. Cambridge: Cambridge University Press.

Wallis, J., & Dollery, B. (2001). Government failure, social capital and the appropriateness of the New Zealand model for public sector reform in developing countries. *World Development, 29*(2), 245–263. https://doi.org/10.1016/S0305-750X(00)00092-9.

Wang, B., & Vongalis-Macrow, A. (2012). Entrepreneurial training project in China: Retraining laid-off workers. *International Education Journal: Comparative Perspectives, 11*(1), 82–95.

Wang, Y. (2011, October 25). *Economic and social status of Chinese women raised—All china women's federation*. Women of China. http://www.womenofchina.cn/womenofchina/html1/13/4737-1.htm.

Winston, C. (2006). *Government failure versus market failure: Microeconomics policy research and government performance*. Washington, DC: AEI-Brookings Joint Center for Regulatory Studies.

Wong, Y. C., & Leung, J. (2012). Long-term care in China: Issues and prospects. *Journal of Gerontological Social Work, 55*(7), 570–586. https://doi.org/10.1080/01634372.2011.650319.

World Health Organization. (2016). *WHO | WHOQOL: Measuring quality of life*. WHO. http://www.who.int/healthinfo/survey/whoqol-qualityoflife/en/.

World Health Organization, & National Health and Family Planning Commission of China. (2016). *China country assessment report on ageing and health*.

Wu, B., Carter, M. W., Goins, R. T., & Cheng, C. (2005). Emerging services for community-based long-term care in urban China: A systematic analysis of Shanghai's community-based agencies. *Journal of Aging & Social Policy, 17*(4), 37–60. https://doi.org/10.1300/J031v17n04_03.

Xie, Y. (2016). Understanding inequality in China. *Chinese Journal of Sociology, 2*(3), 327–347. https://doi.org/10.1177/2057150X16654059.

Yan, D. (2006). *China's Employment Policies and Strategies*. OECG.org.

Yang, J. (2012, October). *Continuity or change? Family structure and its consequences in transitional China*. 10th Rhodes Forum., Austria.

Yang, M. (1994). Reshaping peasant culture and community: Rural industrialization in a Chinese village. *Modern China, 20*(2), 157–179. Retrieved from www.jstor.org/stable/189314.

Ye, M. (2008). *The role of NGO in urban cities: Comparing Shanghai and Taipei*. Shanghai Academy of Social Sciences.

Ye, M. (2010, October 11). *Consumer satisfaction of community services for older adults in Shanghai, China*. The 63rd annual scientific meeting of The Gerontological Society of America, New Orleans, LA.

Ye, M., & Chen, Y. (2014). The influence of domestic living arrangement and neighborhood identity on mental health among urban Chinese elders. *Aging & Mental Health, 18*(1), 40–50. https://doi.org/10.1080/13607863.2013.837142.

Ye, M., Chen, Y., & Peng, Y. (2017). A new era in living arrangements: Determinants of quality of life among Chinese older adults. In M.-C. Tsai & W. Chen

(Eds.), *Family, work and wellbeing in Asia* (pp. 43–64). Springer Singapore. https://doi.org/10.1007/978-981-10-4313-0_3.

Yin, Z. (2016, July 4). *The Shanghai old population analysis*. Sohu.Com. http://news.sohu.com/20160705/n457804985.shtml.

Yin, Z. (2019, October 7). *It is necessary to support elderly care services*. Silver Industry. https://www.silverindustry.cn/pressdetail/1978.html.

Yılmaz, O., & Zeng, Y. (2016). Older parents enjoy better filial piety and care from daughters than sons in China. *American Journal of Medical Research, 3*(1), 244–272.

Zeng, X. (2010). The construction of KPI system for local government performance evaluation. *Journal of Central South University of Forestry & Technology (Social Sciences), 4*. http://en.cnki.com.cn/Article_en/CJFDTotal-ZNLS201004012.htm.

Zeng, Y., Land, K. C., Gu, D., & Wang, Z. (2014). Dynamics of households and living arrangements in the eastern, middle, and western regions of China. In *Household and Living Arrangement Projections* (pp. 225–236). Springer Netherlands. https://doi.org/10.1007/978-90-481-8906-9_13.

Zhan, H. J., & Montgomery, R. J. V. (2003). Gender and elder care in China: The influence of filial piety and structural constraints. *Gender & Society, 17*(2), 209–229. https://doi.org/10.1177/0891243202250734.

Zhang, F., Li, D., Ahrentzen, S., & Zhang, J. (2019). Assessing spatial disparities of accessibility to community-based service resources for Chinese older adults based on travel behavior: A city-wide study of Nanjing, China. *Habitat International*. https://doi.org/10.1016/j.habitatint.2019.05.003.

Zhang G. (2015). *The elderly care service system for older adults in rural areas*. Beijing, China: China Social Sciences Press.

Zhang, J. (2017). The evolution of china's one-child policy and its effects on family outcomes. *Journal of Economic Perspectives, 31*(1), 141–160. https://doi.org/10.1257/jep.31.1.141.

Zhang, L., Curhan, G. C., Hu, F. B., Rimm, E. B., & Forman, J. P. (2011). Association between passive and active smoking and incident type 2 diabetes in women. *Diabetes Care, 34*(4), 892–897. https://doi.org/10.2337/dc10-2087.

Zhao, Y. (2002). Measuring the social capital of laid-off Chinese workers. *Current Sociology, 50*(4), 555–571. https://doi.org/10.1177/0011392102050004005.

Zhou, D. (2010). *Reforming China: Experiences and lessons* (T. Li, Ed.). Singapore: Enrich Professional Publishing.

CHAPTER 4

Geographic Proximity in the Community Eldercare

Drawing on the social-ecological framework, this chapter examines how district government policies influence the HCBS program on the macro level. We employ the geographic information system (GIS) strategy to analyze the patterns of how the District government chose the locations of service centers to expand the HCBS program for older adults. Using government archives, policy documents, and census data, we aim to dissect the strategies for allocating public resources to optimize the HCBS program's ability to accommodate older adults' needs in the old Jing'an District.

4.1 Allocating Health Care in the Community

There are two common ways to make healthcare and related services available for individuals. One way is for various agencies and organizations with flexibility and service specialties to have a customized package but relatively little centralized planning (Anderson & Knickman, 2001; Ekman, Hedman, Swedberg, & Wallengren, 2015; Griffiths, Austin, & Luker, 2004). For example, Ekman and colleagues (2015) reviewed a person-centered care model to provide health care for patients in Sweden. The model involved patients, family caregivers, and multiple healthcare professionals to create a unique healthcare package for the patients. In another study, Griffiths and colleagues (2004) interviewed 8 team members who worked in an interdisciplinary team offering rehabilitation

L. Chen and M. Ye, *Community Eldercare Ecology in China*,
https://doi.org/10.1007/978-981-15-4960-1_4

services for community-dwelling older adults in the United Kingdom. Although both studies found positive results, including reducing the length of hospital stay, reducing patient anxiety and uncertainty (Ekman et al., 2015), and cohesively meeting patients' needs at home (Griffiths et al., 2004), these studies also revealed significant challenges, such as a high demand for professionals (Ekman et al., 2015) and the difficulties of working across large geographic areas and the inflexibilities of traditional professional rules (Griffiths et al., 2004).

Another approach to providing these services is to offer centralized planning to better coordinate different types of care (Fahlström & Kamwendo, 2003; Hickerson et al., 2008; Matsui & Capezuti, 2008). For example, Fahlström and Kamwendo (2003) introduced Sweden's community-based services with the idea that public services and care should be distributed such that every older adult living in the community has guaranteed access to them. One of the advantages of a centralized care plan is the ability to incorporate a wide range of agencies to provide discrete services (e.g., congregate meals, day-care services, or specialists; Fahlström & Kamwendo, 2003; Matsui & Capezuti, 2008).

In the context of HCBS in Shanghai, centralized planning that emphasizes care coordination among discrete services may be a better option because it addresses a wider range of needs (Fahlström & Kamwendo, 2003; Hickerson et al., 2008; Matsui & Capezuti, 2008). Of particular note, HCBS in Shanghai aims to increase service availability and expand service delivery to socioeconomically disadvantaged older adults (Shanghai Civil Affairs Bureau, 2010a).

When public services aim for a broad reception among older adults, they must sometimes turn away from market-oriented strategies (Bhalla, 2001; Hudson, 2015; Winston, 2006). Because the benefits are publicly owned and people can enjoy services without paying additional fees, the private market may show low performance. For example, Hudson (2015) described the collapse of the healthcare system in the United Kingdom, arguing that when outsourced providers are reluctant to service the public, the government should assume responsibility. Market failures may also relate to some endogenous factors of public services because more poor than rich people tend to use the services. Economically disadvantaged populations have fewer personal resources to spend on services than their more affluent counterparts (Andersen & Anderson, 1967), and previous studies in Western contexts have found greater utilization of HCBS programs among disadvantaged and low-income older adults

(Alkema, Reyes, & Wilber, 2006; Borrayo, Salmon, Polivka, & Dunlop, 2002). In light of the growing public need for HCBS programs in China (Y.-M. Chen, Thompson, Berkowitz, Young, & Ward, 2011; Z. Feng, Liu, Guan, & Mor, 2012; Jing An News, 2015; Shanghai Observer, 2018a, 2018b; B. Wu, Carter, Goins, & Cheng, 2005), it is critical to analyze access to and availability of public HCBS resources to guide policy and program development.

Indeed, the characteristics of HCBS programs' locations can reflect many aspects of social justice and inequalities regarding community-dwelling older adults' needs (Pan, Zhao, Wang, & Shi, 2016), demographic distribution (Lee, Shannon, & Brown, 2015), and the public resource allocation process (Abednego & Ogunlana, 2006). In this chapter, we review the strategies for allocating service centers within this centrally designed and administered HCBS program and discuss how these strategies reduce service disparities and influence older adults' ability to age in place in Shanghai. To our knowledge, this work is among the first to explore spatial accessibility to HCBS programs for older adults in urban China, and we hope the results illuminate potentially unmet needs among older adults and guide policymakers in developing future HCBS programs.

Despite the increasing availability of HCBS programs in Shanghai, there is a lack of rigorous analysis regarding their resource allocation. This chapter uses GIS to examine the geographic accessibility (i.e., location patterns) of HCBS service centers. GIS has been widely used to analyze geographic influences on health using a "spatial analysis tradition" (Gatrell & Senior, 1999, p. 925). In addition to modeling spatial disease incidences, GIS is a valuable tool for informing and analyzing healthcare service delivery, such as resource allocation, location, and access to care for different populations (Gatrell & Senior, 1999).

In particular, GIS can help visualize health service accessibility over space and time, allowing for monitoring the development of healthcare services and government policies (Higgs, 2004). Gulliford and colleagues (2002) distinguished public service accessibility based on "having access" and "gaining access." "Having access" relates to service availability for a target population while "gaining access" refers to individuals' ability to overcome barriers and utilize services (Gulliford et al., 2002; Higgs, 2004). Previous studies on public service accessibility have focused primarily on factors related to gaining access, such as affordability (Ensor & Cooper, 2004), mobility (Gulliford et al., 2002), and acceptability

(Jacobs et al., 2012). Meanwhile, limited studies have investigated the use of GIS for "having access." Having access requires an adequate supply of services and allocating the supply geographically to meet community needs (Gulliford et al., 2002).

Specifically, it is crucial to know what factors influence public service location-allocation patterns. Examining service locations, location-allocation strategy concerns the efficient implementation of public services within a bounded space, such as minimizing geographic distance between the target users and their respective centers as well as maximizing coverage for the populations in need (Cromley & McLafferty, 2012; McLafferty, 2003). However, choosing a location for service centers within a limited budget requires striking a balance between cost and public welfare concerns (Teitz, 1968). In Western contexts, many studies on the location allocation of public health services argue for allocating services based on target populations (e.g., people with disabilities), distances between residential areas and the nearest service centers, travel time, and transportation preferences (Harper, Shahani, Gallagher, & Bowie, 2005; Johnson, Gorr, & Roehrig, 2005). Taking these parameters into consideration, location allocation is instrumental in determining how and where services are divvied across regions and communities. As a result, public service location-allocation patterns can reflect social inequality and inequity in meeting individuals' needs (Lane, Sarkies, Martin, & Haines, 2017; Pan et al., 2016).

When investigating public service allocation in China, scholars have identified additional government and cultural characteristics. For example, Pan and colleagues (2016) analyzed the locations of public and private hospitals to understand how public hospitals dominate services in rural China. Z. Wang, Zhang, and Wu (2016) found that cost and geographic distance determined where local governments would build nursing homes. In particular, Chinese older adults requested nursing homes close to their children's homes (Z. Wang et al., 2016). The common requirement of proximity to their children suggests family and family caregiving remain traditional Chinese values. Article 49 of the 1982 Constitution of the People's Republic of China stipulates that parents have the duty to rear and educate their minor children, and adult children have the duty to support and assist their older parents (The National People's Congress of the People's Republic of China, 2000). The 1996 Law of Protection of the Rights and Interests of Older Adults of the People's Republic of China stipulates that adult children are legally obligated to take care of their older parents. Article 11 specifies that adult children pay for their older parents' living expenses, provide instrumental and emotional support, and tend to older parents' needs (The National People's Congress of the People's Republic of China, 1996). Facing increased migration within its borders and recognizing that proximity to their adult children may not be sustained or guarantee family caregiving for older adults—especially the growing population of older adults living alone—the Chinese government amended the Law of Protection of the Rights and Interests of Older Adults of the People's Republic of China in 2013 (The National People's Congress of the People's Republic of China, 2013). This amendment was a policy effort to address the decline in family caregiving for Chinese older adults, which fell from about 15% in 1984 to 12% in 2004 (Houser, Gibson, Redfoot, & AARP Public Policy Institute, 2010). Houser and colleagues (2010) reported that the number of Chinese older adults who live in communities and with disabilities has also increased. Given the discrepancy between older adults' longing for sustainable family caregiving and the lack of proximity to their adult children, HCBS programs may be a feasible solution for providing instrumental and emotional support, regardless of the distance between them and their adult children.

Although HCBS programs in China have rarely been investigated geographically, studies on Chinese HCBS programs have shown the importance of such services for older adults' aging in their current home (Y.-M. Chen et al., 2011). For example, older adults have returned to a community with HCBS services even after temporary hospitalization (Y.-M. Chen et al., 2011). In particular, existing research on HCBS programs in Shanghai has reported their positive associations with older adults' quality of life such as heightened daily activity (Q. Feng et al., 2013) and increased social interactions during mealtime (Ye, Chen, & Kahana, 2016). However, it remains unclear how geographic location and allocation of HCBS programs can influence community-dwelling older adults' life in Shanghai.

4.2 Characteristics of the HCBS Program Area

We examined the spatial planning patterns of the HCBS program in the Jing'an District. As one of the most developed areas of Shanghai in 2013, the old Jing'an District government spent 999 billion RMB in government subsidies (about 145 billion American dollars) on civil affairs, including the HCBS program for elderly residents (Shanghai Bureau of Statistics, 2019).

Originally, the HCBS program in the Jing'an District planned to provide only door-to-door services for older adults who were old-old (Age 80+), low-income, or unable to take care of themselves. As the director of the HCBS program said,

> In 2005, the mainstream idea of taking care of older adults was to build more nursing homes. However, the price of housing in this area was too high so the district government would not have had enough money to build enough nursing homes to house the large number of older adults in the area. As a result, the Department of Aging in the Jing'an District decided to find other ways, such as home care services and some community-based support, to supplement family caregiving. Because the nanny market was chaotic at that time, many older adults told us that they were concerned about asking a nanny to come to their homes to help them. These older adults did not feel safe. We knew that there was a market for providing door-to-door services. We then got financial support from the Department of Aging and recruited some people to start a team to offer home care. We set up a tracking system to manage these door-to-door services. After two years, the team recognized the necessity of setting up a service center in the community as an intersection between older adults' homes and communities, to provide concentrated services for those living nearby. In doing so, this service center can extend coverage to a larger number of older adults, not only for those who are low-income, but any older adult who lives nearby can go to this site to enjoy the service.

Because the HCBS program was available to older adults free of charge, travel time between home and service center was the main cost for community-dwelling older adults in the District. As such, geographic distance and local demographic characteristics are key factors. As the HCBS director continued:

> We located our first pilot service center in one neighborhood in the Jing'an Temple sub-district because of its high concentration of older adults. We also did a small, neighborhood-based survey to estimate the service demands among these older adults. Demand was high across many service items, such personal care, bathing, health promotion, and meal service, which was the highest. So we wanted to experiment with a service center in the neighborhood, and its popularity has continued to grow today.

At the same time, the HCBS program did not forget to provide support for disadvantaged older adults (Shanghai Civil Affairs Bureau, 2010a, 2010b). Achieving service equity became another key mission for the HCBS program, taking older adults' socioeconomic status into consideration. For example, one of the elderly clients, Mrs. Wang, told us:

> A neighbor of mine lives alone after her husband passed away and she has limited means because she does not have children. Staff members at the closest service center regularly visit her and deliver service coupons to her every month. I applaud [the HCBS program's] efforts in supporting these older adults who are in need. This is very important to older adults. We need to know that we are cared for by others and we are not alone.

Also, given the population density and social interactions in China, the physical environment of housing can significantly affect older adults' subjective well-being in China (Ying, Ning, & Xin, 2015). Studies have shown that older adults in China think neighborhood environment is important for their basic needs, physical health, and quality of life (J. Feng, Tang, & Chuai, 2018; Ying et al., 2015). Also, housing conditions are positively associated with social stratification and occupational status among Chinese older adults before they retire (Li & Chen, 2011). This relates to the fact that the physical environment of housing may affect social inequality (Yi & Huang, 2014) and individuals' socioeconomic status (Z. Wang et al., 2016) because "housing is one of the most understudied aspects of aging in China" (Li & Chen, 2011, p. 464). In particular, the current real estate market is a significant revenue source for the Chinese government and its citizens (Besbris & Faber, 2017; J. Chen & Zhang, 2012; Yi & Huang, 2014). Thus, this analysis considers the housing environment (e.g., housing types, age of building) a strong indicator to help understand the allocation patterns of service centers in the HCBS program.

The three most common types of housing in the old Jing'an District are high-end apartments, *Lilong* housing, and *Shikumen* housing. Both *Lilong* and *Shikumen* are old architectural styles that were used in central Shanghai between the late nineteenth and early twentieth centuries (Wan & Ge, 2011; Y. Wang & Chen, 2010; Yun, 2015; J. Zhang & Weng, 2006). *Shikumen* buildings are brick or brick-and-wood mixed structures without modern sanitary facilities, and most of them house low-income families (Guan, 1996). While *Lilong* buildings came later than *Shikumen* and most were designed for middle-class families at the time, these buildings are brick or brick-and-concrete mixed structures with improved interior facilities, such as modern sanitary facilities and larger spaces (Guan, 1996).

Since 1998, when the Chinese government abandoned the old housing system, many old buildings such as *Lilong* or *Shikumen* have been demolished and replaced with high-end apartments that the government then sold (Ren, 2008, 2014; Shen & Wu, 2012; F. Wu, 2016). As a consequence, some sub-districts may have both old housing and high-end apartments. Based on commercial data from the second quarter of 2019 in Shanghai, the average price of a new high-end apartment was about 110,000 RMB per square meter (Savills China, 2019). The average price of an already existing apartment was more than 70,000 RMB per square meter in December 2019, and the highest price was more than 190,000 RMB per square meter (Fang.com, 2019).

We acquired relevant data from the Department of Aging in the Jing'an District, including the geographic and administrative information (2015) of each service center as well as its residential communities' geographic (2007) and demographic data (2015). Information about two residential communities was incomplete, so we analyze the information from 70 residential communities in the five sub-districts.

The geographic and administrative information of each service center includes the address of the service center, the sub-district to which the service center belongs, and the residential communities each center serves. The points on the maps represent the service centers' locations by using coordinate information (latitude and longitude) from their address. The sub-district that the service centers belong to and the residential communities that the service centers are responsible for were also labeled. Each sub-district and residential community were assigned an identification number.

The geographic and demographic information on residential communities was collected in 2015 and included: (a) boundaries of each residential community, (b) the housing type in each residential community, (c) the number of older adults living in the residential community area, and (d) the number of older adults who lived alone. The boundary of each residential community formed a polygon shape on the map of the Jing'an District. Demographic information was coded as continuous variables. The housing types were categorized into three groups: no old housing, *Lilong*, and *Shikumen*.

4.3 Spatial Patterns of Service Centers

Table 4.1 presents the number of service centers in each sub-district. Figure 4.1 shows the geographic locations of all service centers of the HCBS program in the old Jing'an District. Each sub-district has at least six service centers. In particular, sub-districts 03 and 05 each have eight service centers. Geographically speaking, service centers are not dispersed evenly within each sub-district. Our nearest-feature analysis suggests that each residential community is assigned a geographically closest service center. The distance between a residential community and its nearest service center ranges from 0 to 343 meters. On average, the walking distance between a residential community and its closest service center is 44 meters. The proximity was indeed a critical factor for elderly clients to consider. For example, Mrs. Zhu described her views on the service center that she visited daily:

> My own children live far away, in Pudong New Area, which takes about an hour and half to two hours one way to come to see me. They also have their own grandchildren now. I don't want to trouble them much so they don't come to see me very often. Thank God the service center is so close to my home. It only takes 5 minutes for me to get there. Distance matters for older adults. It's not challenging at all. Because I come every day, the service center is like another home for me. These staff members are like my family.

We used GWR4 to test the relationship between the location of each service center and its nearby residential communities' geographic and demographic features (Table 4.2). In Table 4.2, Model 1, spatial logistic

Table 4.1 The number of service centers in each subdistrict in the old Jing'an District

Sub-District	*Number of service centers*
Sub-District 01	6
Sub-District 02	7
Sub-District 03	8
Sub-District 04	7
Sub-District 05	8

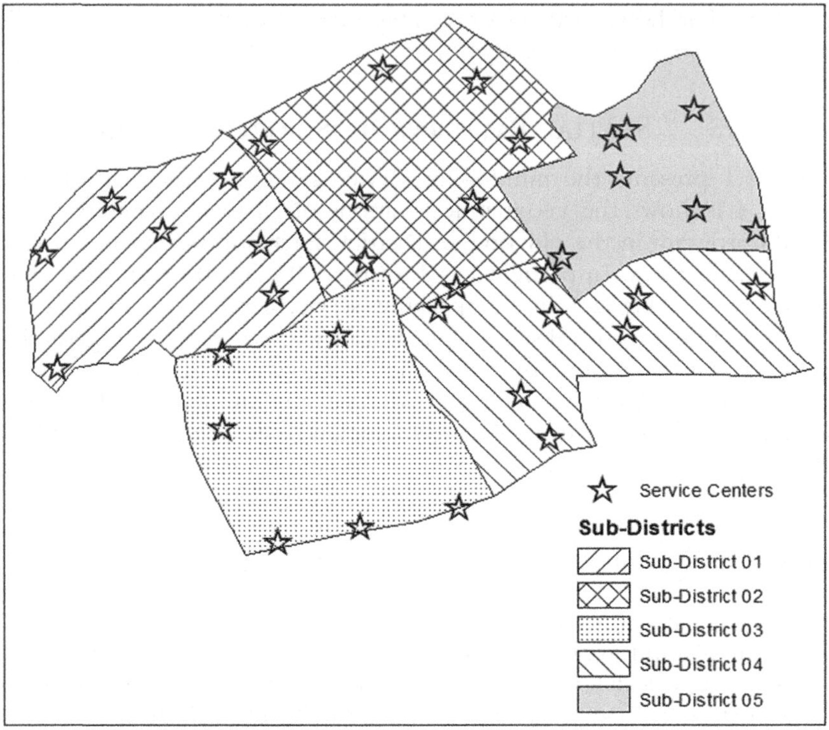

Fig. 4.1 Locations of HCBS service centers in the old Jing'an District

Table 4.2 The relationships between aging population, *Lilong* housing, and service centers

	M1		*M2*		*M3*	
	Estimate	*z*	*Estimate*	*z*	*Estimate*	*z*
Intercept	−1.72	−2.34	−0.49	−1.53	−1.79	−2.38
Aging population	0.00	2.45			0.00	1.97
Lilong housing			1.13	2.24	0.86	1.63
Akaike information criterion (AIC)	95.86		97.36		95.37	

regression analysis identifies a significant relationship between the likelihood of allocating a service center in a residential community and the aging population living in the same residential community ($z = 2.45$). Figure 4.2 shows the locations of service centers in relation to the aging population of a residential community.

Table 4.2, Model 2 tests whether having a service center is related to the primary architectural styles in a residential community. No significant relationship exists between *Shikumen* and the allocation of service centers, nor does such an association exist between old architectural styles (i.e., combining *Shikumen* and *Lilong*) and the allocation of service centers. Interestingly, *Lilong* architecture in a residential community is associated with the allocation of a service center ($z = 2.24$). (See Fig. 4.3.)

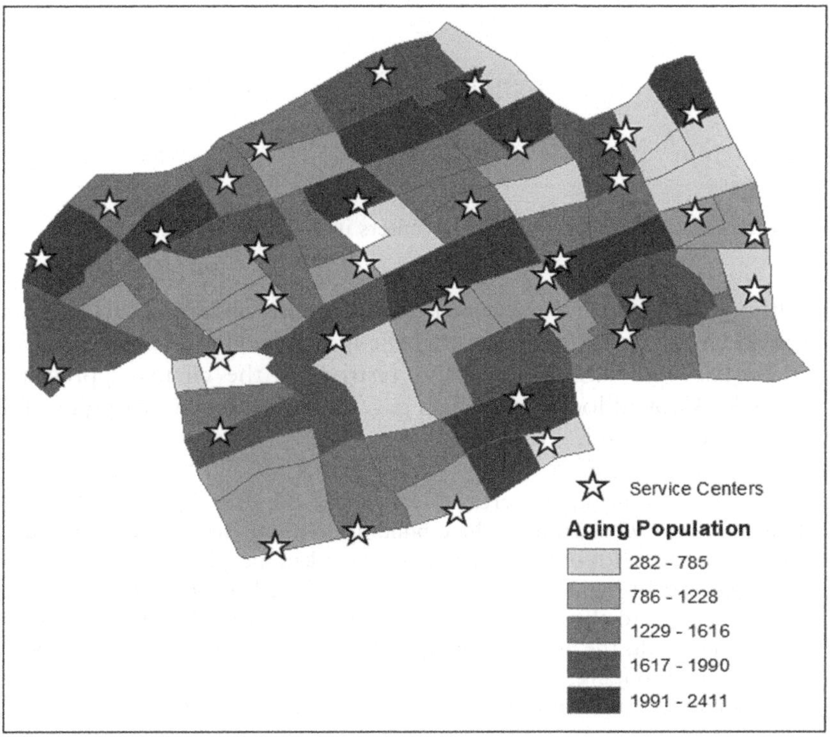

Fig. 4.2 Aging population and service centers in the old Jing'an District

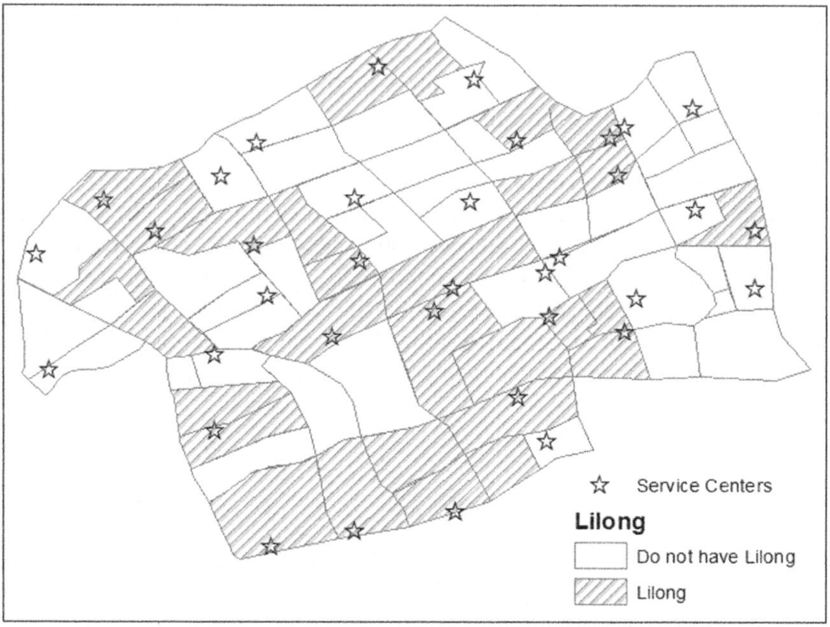

Fig. 4.3 *Lilong* housing and service centers in the old Jing'an District

Although the service centers themselves had different architectural styles and may have posed some challenges for elderly clients, such as limited space for dining areas and restrooms, the clients appreciated these old-fashioned locations, which gave them a sense of belonging. For example, Mr. Hu said,

> I actually grew up in this area. It was about 80 years ago, considering that I am now 93 years old. These buildings did not change much, which brings me back a lot of fond memories. Nowadays, people may think these buildings cannot live but for me, it was one of the best places I have ever lived. I still go to the community library in a similar small building every week. The smell, the atmosphere, and the steep wooden stairs all give me a sense of belonging. This is my home and these are my roots in the community and in Shanghai.

Echoing Mr. Hu's perspective, Ms. Shen, a staff member, shared a similar view:

> Because most elderly clients grew up in the same neighborhood, they are more than familiar with the community and residential committee. Some of them also taught us about the history and locality of this community, which is very useful for us when familiarizing ourselves with the whole neighborhood and providing care. With such ample information, we are able to quickly establish connections with our elderly clients by regular home visits and activity programming in the service center.

When taking both the characteristics of the aging population and architectural styles of the residential community into consideration to predict the likelihood of allocating a service center (Table 4.2, Model 3), the aging population remains significant ($z = 1.97$) while architectural styles become insignificant ($z = 1.63$). However, the Z-score of architectural styles approaches the marginal edge of significance (i.e., $z = 1.645$). This change may be related to the small sample size of service centers. Another possible reason could be the confounded effects between the aging population and the old architectural styles. A residential community's aging population is positively associated with having *Lilong* buildings (not shown, $z = 2.60$). Therefore, the influences of the aging population and the architectural styles may be confounded. In fact, Fig. 4.4 shows that residential communities with larger aging populations and *Lilong* housing are more likely to have a service center allocated. Ms. Pan, one the service center staffs members, commented on the physical environment of the center:

> It is, indeed, not very big for all the elderly clients in this service center… but we try to accommodate them as much as possible. We divide elderly clients into groups to have their lunches and dinners on site. I should admit that it is a bit crowded but we manage. Despite the crowdedness, we are neatly equipped with a restroom with a shower; even disabled elderly clients with wheelchairs can use them. Also, the traditional architecture of our service center attracts elderly clients to hang out here. It is important for them to communicate and socialize with one another as well as with us. The atmosphere is more than caregiving per se, but a sense of belonging and a sense of home. I believe building and its location near the service center has done some trick to attract elderly clients.

By policy design, the HCBS program designates a service center per two to three residential communities. We used the roster to compare

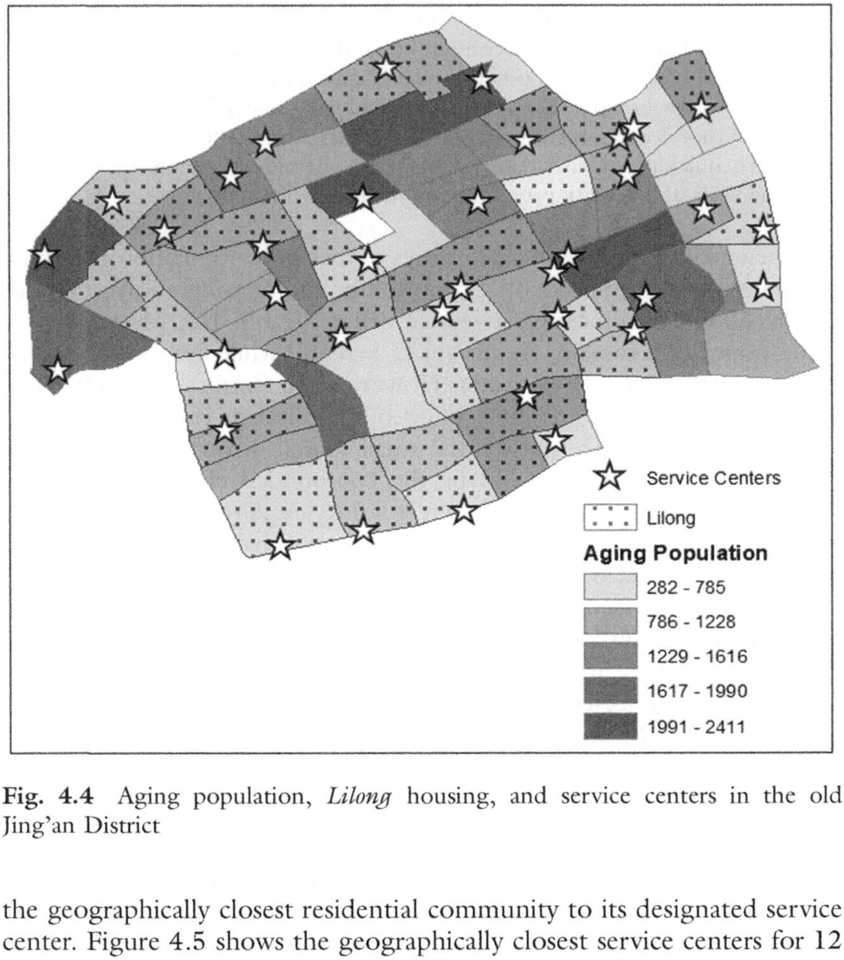

Fig. 4.4 Aging population, *Lilong* housing, and service centers in the old Jing'an District

the geographically closest residential community to its designated service center. Figure 4.5 shows the geographically closest service centers for 12 residential communities, which are different from their assigned centers.

Two allocation discrepancies thus emerge. The first discrepancy is that some of the residential communities and their designated service centers are not located in the same sub-districts. Figure 4.6 shows that residential community A is located in sub-district 01, while its nearest service center is located in sub-district 03. Likewise, residential community B is located in sub-district 05, but its closest service center is located in sub-district 02. This poses a problem because older adults in each residential community must go to the service centers in their own sub-district. One possible

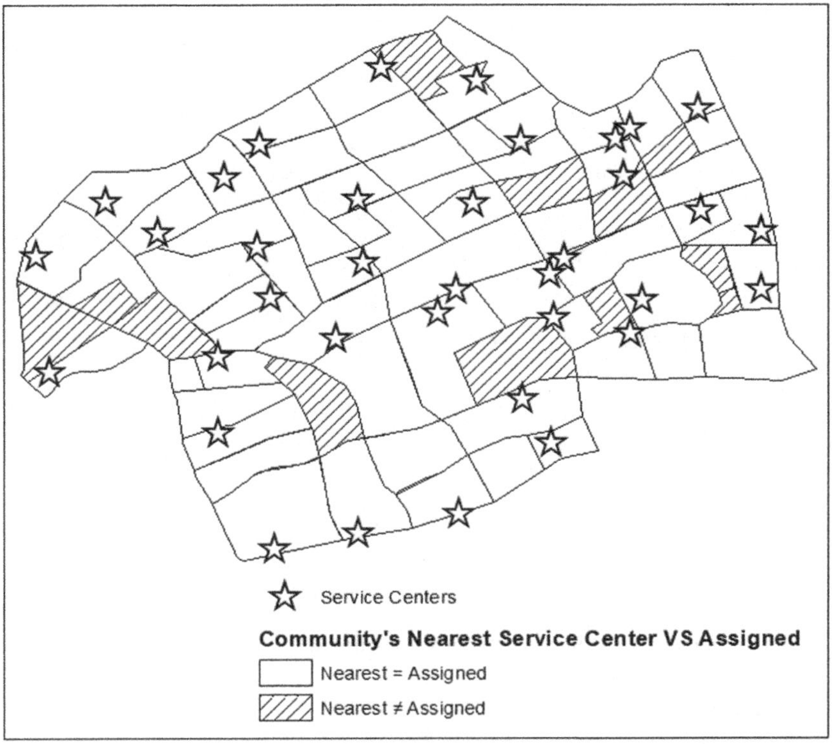

Fig. 4.5 Residential neighborhoods with closest service center other than assigned one

scenario is that older adults are not able to cross a street to the closest service center but must walk for about 10 minutes to the one in their own sub-district.

The second discrepancy is that while designated service centers and geographically closest ones may be located in the same sub-district, they are not the same for 10 residential communities (Fig. 4.7). Most of these communities are in the center of sub-districts, which is far from all the service centers, regardless of designated or nearest ones. Older adults living in these communities have to spend more time walking than those who live along the boundary of a sub-district. As such, the allocation

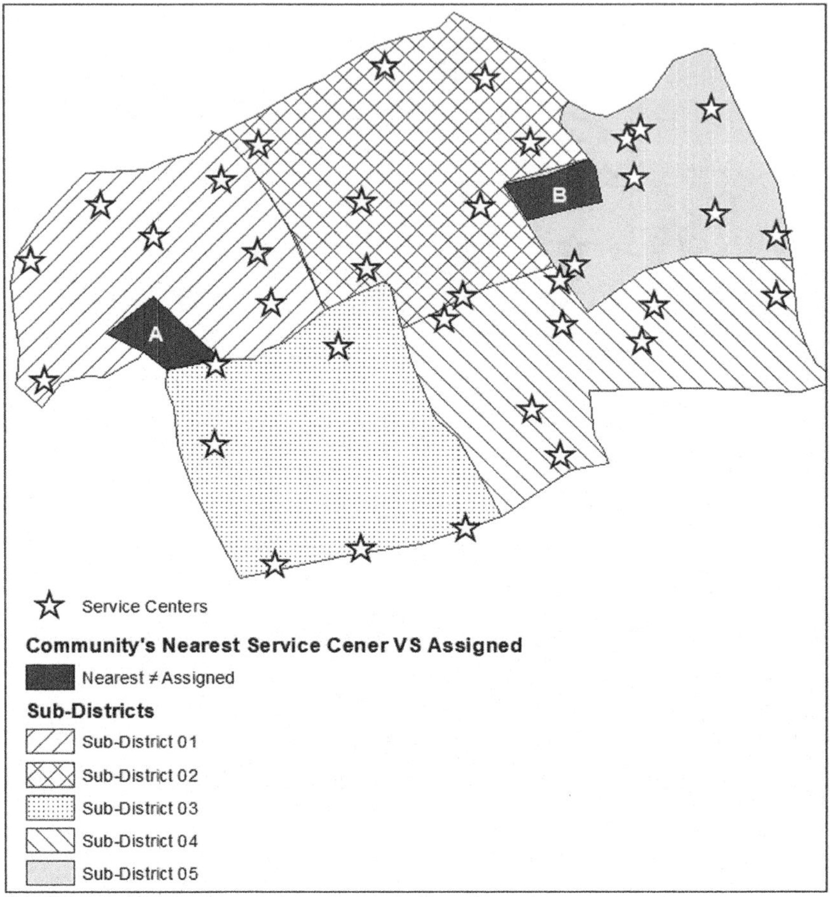

Fig. 4.6 Residential neighborhoods with closest service center other than assigned one across sub-districts

strategy designed by policymakers faces challenges when meeting older adults' needs in real-life scenarios.

Staff members and elderly clients mentioned two discrepancies in service center allocation during our interviews. Many were concerned about the issue of traveling to service centers. For example, Mr. Wei said,

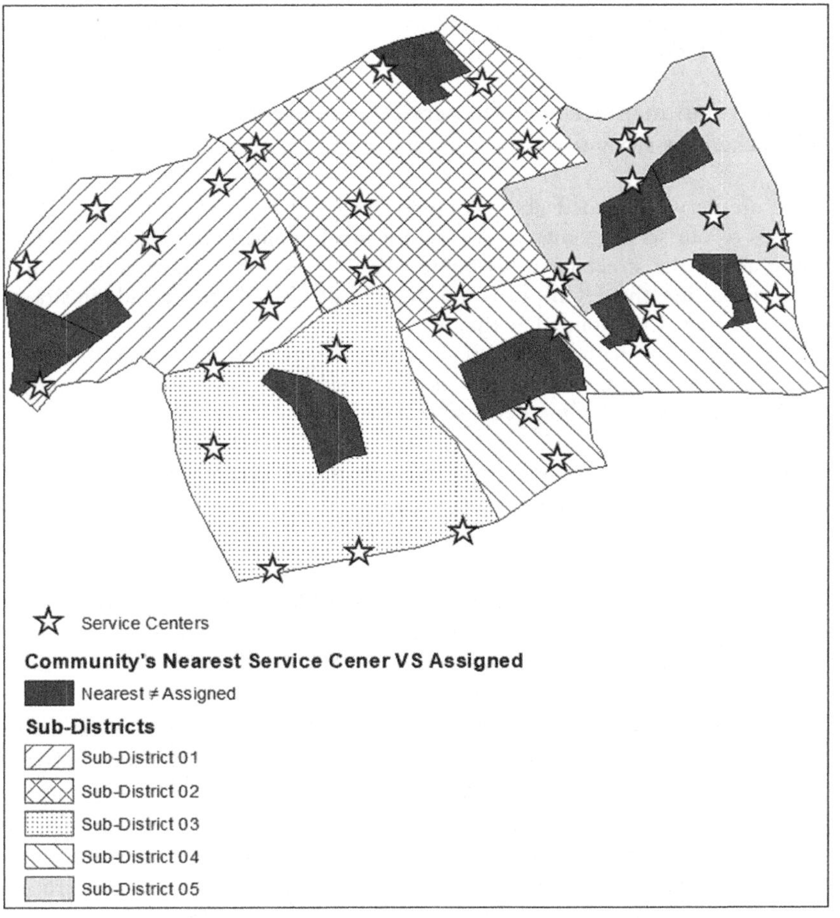

Fig. 4.7 Residential neighborhoods with closest service center other than assigned one within sub-districts

My right leg suffers from chronic pain due to complications from shingles two years ago. I cannot walk for over 10 minutes, otherwise it's going to be excruciating. My designated service center is about 10 minutes away but the closet one is just across the street. I have asked the residential committee and the HCBS program about this issue for many times. They

> always decline my requests, saying these services are allocated based on our household registration status. Such a disappointment!

Ms. Ye, a staff member from the service center where Mr. Wei goes every day, backed his story up:

> We are very concerned about Grandpa Wei. He is in constant pain. He limps to our service center every day for lunch, while there is another one just across the street from his home. We tried to solve this issue for him by contacting the HCBS program board and the sub-district administration people. I'm afraid that they were not of much help. They insist on complying with the household registration system. You are supposed to go to the service center in the sub-district where you household is located. You cannot cross the line between sub-districts...

Structural barriers like this unintentionally posed challenges for frail older adults to conveniently access service centers. Staff members pointed out that when older adults need specific or special support, the line between sub-districts should be dissolved through service and resource allocation.

Taking these two discrepancies into consideration, we controlled for the aging population in each sub-district to further examine the location pattern. Because not all geographically closest service centers are the ones assigned to residential communities, there are two categories of older adults for each service center: those who live close enough and those who are assigned to come to the service center. The largest difference between the numbers of these two groups of elderly is about 2200 older adults. Figure 4.8 uses arrows to illustrate which service center is geographically closest (blue) for the residential communities and which is the designated center (pink) in the 10 residential communities in Fig. 4.7. Figure 4.8 reveals that all the geographically closest service centers for the 10 residential communities serve a larger number of older adults living nearby than the designated service centers do. Possible explanations include that the capacity for serving older adults among these 10 service centers may be larger than the 10 designated ones, or the physical space of these 10 service centers may be larger than the 10 designated ones, or the quality of service of these 10 service centers may be better than the 10 designated ones. Regardless, these 10 service centers attract more elderly clients than they are designed to accommodate.

In our interviews, staff members also found these discrepancies interesting. For example, Ms. Sun said,

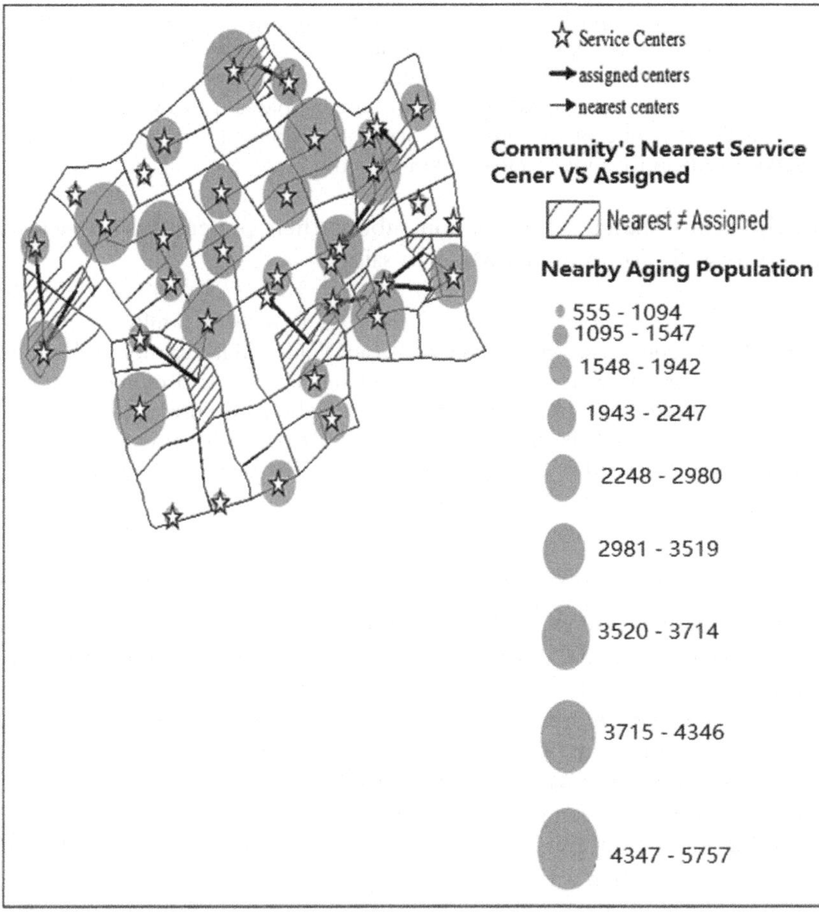

Fig. 4.8 Residential neighborhoods with closest service center other than assigned one within sub-district, possibly because of center service capacity

It is true that in our sub-district, some service centers attract more elderly clients while some do not. The HCBS program tries to even out the differences in service delivery and quality of care by training us and making strict regulations. Still, elderly clients show their preferences. When we see these, first we tell elderly clients to choose the service center closest to their homes, for safety purposes. Second, we tell them about the capacity

> of each service center to give them an understanding of what they should expect. If these two reasons cannot persuade the elderly client, we just let them stay where they prefer. I have noticed, though, that when a group of elderly clients become familiar, they are inseparable and do want to hang out daily. It is one of the most important functions of community-based service centers for older adults. I think it's the best.

Elderly clients shared their opinions about their preferences for service centers as well. For example, Mr. Bao said,

> The closer the service center is, the better, for most older adults receiving the HCBS program services. It is true because older adults cannot walk too far. If it takes more than 15 minutes walking, I feel it is tiresome. If the service center is close by, then many of my neighbors go there. So there is the familiarity, which is an additional incentive to go. I have regular poker games with three other elderly clients living in the same community. The service center gives us the place to play, which is wonderful.

4.4 Service Center Location-Allocation Strategy

Given the relatively dense aging population in the old Jing'an District, the HCBS program has gradually allocated 36 service centers across five sub-districts, with at least six service centers each since 2007. Some centers are located in the central areas of sub-districts while others are located along sub-district boundaries. On average, one service center per 0.2 square kilometers is intended to serve older adults.

Our GIS analysis suggests that the size of the aging population in the residential communities is a major determinant for allocating HCBS service centers. When a residential community has a larger aging population, it is more likely to be allocated a service center. Because it receives district and sub-district subsidies, the HCBS program can be viewed as a public service that should be open to the public to meet their welfare needs. In line with the nature of public services, choices of locations for these service centers should incorporate both older adults' needs (e.g., walkable; F. Zhang, Ahrentzen, & Zhang, 2019) and market price (e.g., budget to build and maintain a service center). Market research can shed some light on these financial implications. For example, Wrigley (2014) pointed out that a typical grocery store would be built about 10-minutes driving distance from its potential customers in the United Kingdom to balance its maintenance expenditure and profits.

Simkin (1989) revealed that shopping centers that are easily accessible (e.g., close to bus stations) have the greatest chance to make profits. Girth, Hefetz, Johnston, and Warner (2012) argued that tangible investment and administrative resources are essential factors when managing public services in noncompetitive markets. Thus, financial factors are important not only in the for-profit sector, but also for nonprofit HCBS programs (Girth et al. 2012).

In addition to financial factors, allocating service centers in the residential communities based on the size of their aging populations should also help to decrease older adults' travel time and increase service coverage. Johnson and colleagues (2005) found that extending service distance for a senior center in the United States tended to decrease elderly clients' usage rate. Older adults in Nanjing were not likely to be attracted to HCBS programs in underdeveloped areas (F. Zhang et al., 2019). Thus, allocating the 36 old Jing'an service centers based on the size of their aging population should help older adults access both services (Gulliford et al., 2002) and make their service and service items more attractive (F. Zhang et al., 2019). The HCBS program made sure that the accessibility of each service center was meaningful to older adults living in the neighborhood designated to individual service center. For example, Mrs. Fang described her experience signing up for the HCBS program:

> When I knew the service center was open in our neighborhood, I was very excited. I knew the sub-district government had planned to set up one nearby. I used to go to another one farther away, but it was not convenient for me in terms of distance and social network. I have friends in this neighborhood but I don't know many people there. My friends prefer to not take the long walk to go there. So when the service center opened, we were all eager to go and hang out there.

Indeed, the "HCBS might support older adults with specific characteristics to… stay in their own communities longer" (Y.-M. Chen et al., 2011, p. 376). Proximally located service centers become a convenient social venue for elderly clients, extending their interactions beyond passively receiving care from staff members and other instrumental support. This creates a friendly atmosphere in the service center, which attracts more older adults to such services.

Our analysis also suggests that architectural styles in the residential communities are related to service centers' location-allocation patterns.

Considering the high real estate prices in the old Jing'an District, converting older housing into service centers can save public funds (Colliers International, 2015). In practice, however, old architecture styles (i.e., combining *Lilong* and *Shikumen*) and *Shikumen* alone are not related to service centers' geographical allocation. However, after separating *Shikumen* from *Lilong*, we found that compared to other architecture styles, only those communities where *Lilong* housing prevails are more likely to have a service center allocated.

Compared to *Shikumen*, *Lilong* architecture has greater advantages (Arkaraprasertkul, 2009; Guan, 1996) for serving as an HCBS service center. *Lilong* was introduced to Shanghai later than the *Shikumen* style; and was designed with more advanced technology (e.g., modern sanitary facilities) for upper middle- and middle-class families in the 1920s (Arkaraprasertkul, 2009; Guan, 1996; Triolo, 2019). The mixed brick-and-concrete construction of *Lilong* housing is more stable than the mixed brick-and-wood construction of the earlier *Shikumen* style. With larger windows, *Lilong* interiors are brighter and more spacious than *Shikumen*. On the other hand, people who live in *Shikumen* buildings usually need to go to public kitchen, bathroom, or storage rooms (Triolo, 2019). Life can be very "disordered" when living in a *Shikumen* building (Triolo, 2019, p. 1). With the advent of *Lilong* housing, cars also became popular in Shanghai, since there was a larger adjunct area for parking than for *Shikumen* (Arkaraprasertkul, 2009; Guan, 1996). This spatial advantage offers an HCBS service center with an extra lane for loading and unloading.

Individual modern sanitary system in *Lilong* housing is another advantage for it to be modified into a service center for the HCBS program. Residents in *Shikumen* housing have to share public toilets. It is crucial for a service center to be equipped with a restroom with a shower for older adults' convenience. The renovation for a service center would cost less for the HCBS program when the service center is originally *Lilong* housing rather than *Shikumen*. As such, when deciding a location for a service center, *Lilong*-style housing becomes a better choice than other old architecture styles because of convenient facilities and low renovation costs.

Another location-allocation strategy for establishing a service center for the HCBS program is related to housing types, which to some degree reflect individuals' socioeconomic status (Z. Wang et al., 2016; Yi & Huang, 2014); residential communities with larger low-income aging populations are more likely to be allocated a service center. This finding supports the goal set by the Shanghai government that HCBS programs should promote social equality through affordable and accessible services aimed at disadvantaged older adults (Shanghai Observer, 2018b).

Our analysis reveals that not all the geographically closest service centers are the ones designated by the HCBS program to residential communities. One possible reason is related to the administration and funding of the HCBS program. Although the overall HCBS program is budgeted and subsidized on the district level, service centers are administered and managed on the sub-district level. Each sub-district government autonomously runs the service centers located within its area. So services offered vary across the five sub-districts, though they fall into the same 10 service categories. When older adults live in a community located between two sub-districts, they may have to go to the designated service center in their own sub-district, even though the geographically closest service center may be located across the street in the other sub-district.

Within the same sub-district, some designated service centers are not the geographically closest ones for certain residential communities. In fact, some service centers attract a larger aging population than others simply because they are closer to where older adults live. This phenomenon suggests that older adults usually prefer to go to the geographically closest service centers. However, the workloads for these centers may become overwhelming and force older adults in these communities to go to elsewhere.

This creates a location-allocation discrepancy between the geographically closest service centers and the designated service centers, which can lead to several consequences. First, this discrepancy prevents older adults from gaining access to services. Some have to walk further to access the designated service center rather than to the geographically closest ones. Although the average walking distance between a residential community and a service center is less than 0.5 km, it may still pose some difficulties for frail older adults to walk this distance, especially when they have mobility problems or the weather is not good.

Second, this discrepancy challenges service centers' capacities. Given older adults' geographic preferences, some service centers may serve a larger aging population that exceeds their current capacity, while others serve smaller numbers of older adults than they are supposed to help. In the current HCBS program, the largest difference in the number of older adults for the geographically closest service centers versus the designated one is 2200. This disparity calls for an adjustment to better disperse healthcare resources in these communities and balance older adults' needs with the service centers' capacities.

4.5 Macro- and Meso-Level Meanings of Service Center Locations

In this section, we theorize the relationship between local policy and proximal programs (Greenfield, 2012) to advance understanding of the macro- and meso-level meanings of service center locations and their implications for older adults in the community. Social environments, such as school administrative units, community-based long-term care systems, transportation, or healthcare systems, can have both direct and indirect influences on people's lives (Garza, 2018). In this chapter, we consider the HCBS program as a meso-level construct, while its administration and related policy implementation function as a macrolevel construct in the community ecology.

HCBS policy implementation and program administration can have an impact on older adults' preferences regarding and willingness to utilize HCBS services. As mentioned earlier in the chapter, the HCBS program in the Jing'an District began with door-to-door home care services without a physical service center location. Despite its efficiency, service remained limited to a small group of older adults in one neighborhood. In order to provide centralized services for the growing number of older adults in the community and meet their rising needs, the district government launched physical service centers to expand the accessibility of this HCBS program (Gerontological Society of Shanghai, 2009).

Investing government effort and mandating the allocation of public resources for the HCBS is a way to help balance inequalities among older adults with different socioeconomic backgrounds, material support, caregiving resources, and other advantages and disadvantages (Lane et al., 2017). Increasing the availability and accessibility of HCBS programs can improve the overall well-being and quality of life for older adults (Johnson et al., 2005). Chinese scholars agree that government and not-for-profit organizations should be responsible for increasing service coverage and accessibility for older adults who live in underdeveloped areas and have limited means (F. Zhang et al., 2019). While HCBS programs can contribute to the public good, more strategic planning is required on the macro level to allocate them effectively.

Because geographical preference influences older adults' utilization of HCBS services, strategically allocating HCBS service centers is important. The Jing'an District government requires that the service center be as close as possible to the residential neighborhoods to ensure the convenient access (Gerontological Society of Shanghai, 2009). Recently, the district government has further mandated that every 10,000 older adults have one HCBS service center to establish a "15-minute comfort zone" in their immediate neighborhood (Hou & Liu, 2017). Our GIS analysis shows that the distance between any residential neighborhood and its nearest HCBS service center is within 500 meters. The geographic proximity between older adults' homes and the service centers ensures convenient access to services and encourages service utilization.

Government financial support is essential for allocating HCBS service centers. The director of the old Jing'an District HCBS program pointed out that it was to convert *Shikumen*-style buildings into service centers at a reasonable price. Our analysis also suggests that a *Lilong*-style building is more likely to be turned into an HCBS service center (Guan, 1996). *Shikumen* buildings are much older than *Lilong* ones and with limited built-in facilities. For example, the *Shikumen* buildings do not typically have toilets. However, toilets are necessary for a HCBS service center to be in actual operation. Building new sanitary equipment and facilities require additional financial and workforce support.

In reality, though, older adults who live in *Shikumen* buildings in the Jing'an District generally have greater needs for HCBS services than their better-off counterparts because of their disadvantaged socioeconomic backgrounds. For example, bathing in the winter remains problematic and cumbersome for older adults living in *Shikumen* or *Lilong* buildings (Shanghai Observer, 2018b). Although the Shanghai Municipal Government is actively transforming old buildings so that residents can enjoy the convenience of private bathrooms, there are still many residents who do not have private bathrooms in their homes and need to go out to public restrooms (Hou & Liu, 2017; Kano, Rosenberg, & Dalton, 2018; Shanghai Observer, 2018b). Older adults with limited means and difficult living environments are more in need of HCBS services.

Given the limited workforce of the HCBS service centers in the Jing'an District, it is difficult for each of them to meet all of their clients' needs, especially when there is a large number of older adults living in the area, while the designated service center is small. Staff members mentioned the unbalanced workload among service centers, which challenged their performance and quality of care. Although the HCBS program enjoyed much freedom in the district and sub-districts administer the center autonomously, bureaucratic challenges and discrepancies still exist in the current HCBS program. Older adults living on the borderlines between sub-districts are sometimes perplexed about which service center they should use. They may prefer a service center that is closer distance but in another sub-district. Under current HCBS regulations in the Jing'an District, these older adults must go to the service center in their own sub-district. Regardless of distance, allocating HCBS resources across sub-districts and navigating bureaucratic hierarchies remains a challenge in the Jing'an District.

4.6 Chapter Summary

In this chapter, we have evaluated spatial allocation patterns of the current HCBS program, illuminating its operation on the macro level in the community. We have presented the geographic and demographic features of the 36 service centers of the HCBS program across five sub-districts and 72 residential communities in the old Jing'an District. Employing GWR4 and nearest-feature analysis strategies, we have analyzed the correlation between the locations of the 36 service centers and the residential communities in which they are located. Our analysis reveals that existing service centers of the HCBS program cover most residential communities in the old Jing'an District. In particular, communities with larger aging populations and older housing are more likely to have a service center than other residential communities. We have discussed the location-allocation strategy applied in the HCBS program and balancing older adults' needs and service centers' capacities when allocating healthcare resources at the macro level. In order to achieve extended service coverage while amending service disparity, the government should ensure the greatest possible access for older adults at the beginning of allocation. Strategic allocation on the macro level also requires policymakers to consider both external and internal characteristics of an HCBS.

References

Abednego, M. P., & Ogunlana, S. O. (2006). Good project governance for proper risk allocation in public–private partnerships in Indonesia. *International Journal of Project Management, 24*(7), 622–634. https://doi.org/10.1016/j.ijproman.2006.07.010.

Alkema, G. E., Reyes, J. Y., & Wilber, K. H. (2006). Characteristics associated with home- and community-based service utilization for medicare managed care consumers. *The Gerontologist, 46*(2), 173–182. https://doi.org/10.1093/geront/46.2.173.

Andersen, R., & Anderson, O. W. (1967). *A decade of health services: Social survey trends in use and expenditure.* Retrieved from https://www.cabdirect.org/cabdirect/abstract/19692703151.

Anderson, G., & Knickman, J. R. (2001). Changing the chronic care system to meet people's needs. *Health Affairs, 20*(6), 146–160.

Arkaraprasertkul, N. (2009). Towards modern urban housing: Redefining Shanghai's lilong. *Journal of Urbanism: International Research on Placemaking and Urban Sustainability, 2*(1), 11–29. https://doi.org/10.1080/17549170902833816.

Besbris, M., & Faber, J. W. (2017). Investigating the relationship between real estate agents, segregation, and house prices: Steering and upselling in New York state. *Sociological Forum, 32*(4), 850–873. https://doi.org/10.1111/socf.12378.

Bhalla, A. (2001). *Market or government failures? An Asian perspective.* New York, NY: Springer.

Borrayo, E. A., Salmon, J. R., Polivka, L., & Dunlop, B. D. (2002). Utilization across the continuum of long-term care services. *The Gerontologist, 42*(5), 603–612. https://doi.org/10.1093/geront/42.5.603.

Chen, Y.-M., Thompson, E. A., Berkowitz, B., Young, H. M., & Ward, D. (2011). Factors and home- and community-based services (HCBS) that predict older adults' residential transitions. *Journal of Service Science and Management, 4*(3), 368. https://doi.org/10.4236/jssm.2011.43043.

Chen, J., & Zhang, X. (2012). Housing demand and housing policy of the young sandwich group in mainland china—A survey of young white-collar employees in Shanghai. *Journal of Youth Studies (10297847), 15*(1), 130–143.

Colliers International. (2015). *Shanghai high-end residential market stable despite HPRs.* Retrieved from http://www.colliers.com/-/media/D1B5A256C6BF4DDE972586145260C5FC.ashx?la=en-GB.

Cromley, E. K., & McLafferty, S. (2012). *GIS and public health* (2nd ed.). New York, NY: Guilford Press.

Ekman, I., Hedman, H., Swedberg, K., & Wallengren, C. (2015, February 10). Commentary: Swedish initiative on person centred care. *BMJ*, *350*(14), h160–h160. https://doi.org/10.1136/bmj.h160.

Ensor, T., & Cooper, S. (2004). Overcoming barriers to health service access: Influencing the demand side. *Health Policy and Planning, 19*(2), 69–79. https://doi.org/10.1093/heapol/czh009.

Fahlström, G., & Kamwendo, K. (2003). Increased physiotherapy in sheltered housing in Sweden: A study of structure and process in elderly care. *Health and Social Care in the Community, 11*(6), 470–476. https://doi.org/10.1046/j.1365-2524.2003.00452.x.

Fang.com. (2019, December 31). *Shanghai Jing'an enters "70,000+", with the average price of the most expensive community exceeding 190,000/sq.m.* Retrieved from https://sh.esf.fang.com/newsecond/news/34381720.htm [Chinese].

Feng, Z., Liu, C., Guan, X., & Mor, V. (2012). China's rapidly aging population creates policy challenges in shaping a viable long-term care system. *Health Affairs*, 2764–2773. https://doi.org/10.1377/hlthaff.2012.0535.

Feng, J., Tang, S., & Chuai, X. (2018). The impact of neighbourhood environments on quality of life of elderly people: Evidence from Nanjing, China. *Urban Studies, 55*(9), 2020–2039. https://doi.org/10.1177/0042098017702827.

Feng, Q., Zhen, Z., Gu, D., Wu, B., Duncan, P. W., & Purser, J. L. (2013). Trends in ADL and IADL disability in community-dwelling older adults in Shanghai, China, 1998–2008. *The Journals of Gerontology Series B: Psychological Sciences and Social Sciences, 68*(3), 476–485. https://doi.org/10.1093/geronb/gbt012.

Gatrell, A., & Senior, M. (1999). Health and health care applications. In *Geographical information systems* (p. 19). New York: Wiley.

Garza, L. (2018). *Understanding barriers affecting elders' low utilization of HCBS: A multiple case study of HCBS social workers*. Doctoral dissertation, Northcentral University.

Gerontological Society of Shanghai. (2009). *New community-based long-term care program for older adults: Leling in Jing'an district*. Retrieved from http://www.shanghaigss.org.cn/news_view.asp?newsid=6440 [Chinese].

Girth, A. M., Hefetz, A., Johnston, J. M., & Warner, M. E. (2012). Outsourcing public service delivery: Management responses in noncompetitive markets. *Public Administration Review, 72*(6), 887–900. https://doi.org/10.1111/j.1540-6210.2012.02596.x.

Greenfield, E. A. (2012). Using ecological frameworks to advance a field of research, practice, and policy on aging-in-place initiatives. *The Gerontologist, 52*(1), 1–12. https://doi.org/10.1093/geront/gnr108.

Griffiths, J., Austin, L., & Luker, K. (2004). Interdisciplinary teamwork in the community rehabilitation of older adults: An example of flexible working in primary care. *Primary Health Care Research & Development, 5*(3), 228–239. https://doi.org/10.1191/1463423604pc202oa.

Guan, Q. (1996). *Lilong housing: A traditional settlement form*. https://central.bac-lac.gc.ca/.item?id=TC-QMM-27475&op=pdf&app=Library.

Gulliford, M., Figueroa-Munoz, J., Morgan, M., Hughes, D., Gibson, B., Beech, R., & Hudson, M. (2002). What does "access to health care" mean? *Journal of Health Services Research & Policy, 7*(3), 186–188. https://doi.org/10.1258/135581902760082517.

Harper, P. R., Shahani, A. K., Gallagher, J. E., & Bowie, C. (2005). Planning health services with explicit geographical considerations: A stochastic location–allocation approach. *Omega, 33*(2), 141–152. https://doi.org/10.1016/j.omega.2004.03.011.

Hickerson, B., Moore, A., Oakleaf, L., Edwards, M., James, P. A., Swanson, J., & Henderson, K. A. (2008). The role of a senior center in promoting physical activity for older adults. *Journal of Park and Recreation Administration, 26*(1). https://js.sagamorepub.com/jpra/article/view/1332.

Higgs, G. (2004). A literature review of the use of GIS-based measures of access to health care services. *Health Services and Outcomes Research Methodology, 5*(2), 119–139. https://doi.org/10.1007/s10742-005-4304-7.

Hou, L., & Liu, Y. (2017). Life circle construction in China under the idea of collaborative governance: A comparative study of Beijing, Shanghai and Guangzhou. *Geographical Review of Japan Series B, 90*(1), 2–16. https://doi.org/10.4157/geogrevjapanb.90.2.

Houser, A., Gibson, M. J., & Redfoot, D. L. (2010). *Trends in family caregiving and paid home care for older people with disabilities in the community: Data from the national long-term care survey*. AARP Public Policy Institute.

Hudson, B. (2015). Dealing with market failure: A new dilemma in UK health and social care policy? *Critical Social Policy, 35*(2), 281–292. https://doi.org/10.1177/0261018314563037.

Jacobs, J. A., Clayton, P. F., Dove, C., Funchess, T., Jones, E., Perveen, G., … Deshpande, A. D. (2012). A survey tool for measuring evidence-based decision making capacity in public health agencies. *BMC Health Services Research, 12*(1), 57.

Jing An News. (2015, September 29). Community-based elderly care services in Summer. *Jing An News*. Retrieved from http://shuzibao.jingan.gov.cn/html/2015-09/29/content_327258.htm.

Johnson, M. P., Gorr, W. L., & Roehrig, S. (2005). Location of service facilities for the elderly. *Annals of Operations Research, 136*(1), 329–349. https://doi.org/10.1007/s10479-005-2062-0.

Kano, M., Rosenberg, P. E., & Dalton, S. D. (2018). A global pilot study of age-friendly city indicators. *Social Indicators Research, 138*(3), 1205–1227. https://doi.org/10.1007/s11205-017-1680-7.

Lane, H., Sarkies, M., Martin, J., & Haines, T. (2017). Equity in healthcare resource allocation decision making: A systematic review. *Social Science and Medicine, 175*, 11–27. https://doi.org/10.1016/j.socscimed.2016.12.012.

Lee, J. S., Shannon, J., & Brown, A. (2015). Characteristics of older Georgians receiving older Americans act nutrition program services and other home- and community-based services: Findings from the Georgia aging information management system. *Journal of Nutrition in Gerontology and Geriatrics, 34*(2), 168–188. https://doi.org/10.1080/21551197.2015.1031595.

Li, B., & Chen, S. (2011). Aging, living arrangements, and housing in China. *Ageing International, 36*(4), 463–474. https://doi.org/10.1007/s12126-010-9094-8.

Matsui, M., & Capezuti, E. (2008). Perceived autonomy and self-care resources among senior center users. *Geriatric Nursing, 29*(2), 141–147. https://doi.org/10.1016/j.gerinurse.2007.09.001.

McLafferty, S. L. (2003). GIS and health care. *Annual Review of Public Health, 24*(1), 25–42. https://doi.org/10.1146/annurev.publhealth.24.012902.141012.

Pan, J., Zhao, H., Wang, X., & Shi, X. (2016). Assessing spatial access to public and private hospitals in Sichuan, China: The influence of the private sector on the healthcare geography in China. *Social Science and Medicine, 170*, 35–45. https://doi.org/10.1016/j.socscimed.2016.09.042.

Ren, X. (2008). Forward to the past: Historical preservation in globalizing Shanghai. *City & Community, 7*(1), 23–43. https://doi.org/10.1111/j.1540-6040.2007.00239.x.

Ren, X. (2014). The political economy of urban ruins: Redeveloping Shanghai. *International Journal of Urban and Regional Research, 38*(3), 1081–1091. https://doi.org/10.1111/1468-2427.12119.

Savills China. (2019). *Savills china | Savills Q2/2019 Shanghai property market overview*. Retrieved from https://en.savills.com.cn/insight-and-opinion/savills-news/175991/savills-q2-2019-shanghai-property-market-overview.

Shanghai Bureau of Statistics. (2019). *Shanghai statistical yearbook 2018*. Retrieved from http://tjj.sh.gov.cn/html/sjfb/201901/1003014.html [Chinese].

Shanghai Civil Affairs Bureau. (2010a). *Explanation on the program of the elderly home care in community in Shanghai*. Retrieved from http://www.shmzj.gov.cn/gb/shmzj/node8/node890/userobject1ai27337.html [Chinese].

Shanghai Civil Affairs Bureau. (2010b). *Shanghai civil affairs bureau-elderly home care in community regulations*. Retrieved from http://mzj.sh.gov.cn/gb/shmzj/node8/node15/node55/node230/node246/userobject1ai25433.html [Chinese].

Shanghai Observer. (2018a, October 16). *There are currently 2281 elderly people aged 100 and above in Shanghai. Last year, there were 14.9 centenarians per*

100,000 population. Retrieved from http://www.shanghai.gov.cn/nw2/nw2314/nw2315/nw4411/u21aw1344627.html [Chinese].

Shanghai Observer. (2018b, December 19). *Those who bathed in the "Old Town" in Shanghai*. Retrieved from https://www.shobserver.com/news/detail?id=122558 [Chinese].

Shen, J., & Wu, F. (2012). Restless urban landscapes in China: A case study of three projects in Shanghai. *Journal of Urban Affairs, 34*(3), 255–277. https://doi.org/10.1111/j.1467-9906.2011.00577.x.

Simkin, L. (1989). SLAM: Store location assessment model—Theory and practice. *Omega, 17*(1), 53–58. https://doi.org/10.1016/0305-0483(89)90020-0.

Teitz, M. B. (1968). Toward a theory of urban public facility location. *Papers in Regional Science, 21*(1), 35–51.

The National People's Congress of the People's Republic of China. (1996). *1996 The law of protection of the rights and interests of older adults of the People's Republic of China*. Retrieved from http://www.npc.gov.cn/zgrdw/wxzl/gongbao/1996-08/29/content_1479994.htm.

The National People's Congress of the People's Republic of China. (2000). *1982 Constitution of the People's Republic of China*. Retrieved from http://www.npc.gov.cn/wxzl/wxzl/2000-12/06/content_4421.htm.

The National People's Congress of the People's Republic of China. (2013). *2013 Law of Protection of the Rights and Interests of Older Adults of the People's Republic of China*. Retrieved from http://www.gov.cn/flfg/2012-12/28/content_2305570.htm.

Triolo, L. (2019). *Shikumen: An interior design proposal for limited spaces in traditional dwellings of Shanghai* [Politecnico di Milano]. Retrieved from http://hdl.handle.net/10589/147651.

Wan, Y., & Ge, J. (2011). The conservation and renewal of shanghai shikumen building: Current situation and suggestions. *Fudan Journal (Social Sciences Edition), 4*. http://en.cnki.com.cn/Article_en/CJFDTotal-FDDX201104009.htm.

Wang, Y., & Chen, B. (2010). The fire-fighting design in the Shikumen building reconstruction protection in Shanghai. *Shanxi Architecture, 1*. http://en.cnki.com.cn/Article_en/CJFDTotal-JZSX201001011.htm.

Wang, Z., Zhang, F., & Wu, F. (2016). Intergroup neighbouring in urban China: Implications for the social integration of migrants. *Urban Studies, 53*(4), 651–668. https://doi.org/10.1177/0042098014568068.

Winston, C. (2006). *Government failure versus market failure: Microeconomics policy research and government performance*. Washington, DC: AEI-Brookings Joint Center for Regulatory Studies.

Wrigley, N. (2014). *Store choice, store location and market analysis (Routledge revivals)*. Routledge. https://doi.org/10.4324/9781315736686.

Wu, F. (2016). State dominance in urban redevelopment: Beyond gentrification in urban China. *Urban Affairs Review, 52*(5), 631–658. https://doi.org/10.1177/1078087415612930.

Wu, B., Carter, M. W., Goins, R. T., & Cheng, C. (2005). Emerging services for community-based long-term care in urban China: A systematic analysis of shanghai's community-based agencies. *Journal of Aging & Social Policy, 17*(4), 37–60. https://doi.org/10.1300/J031v17n04_03.

Ye, M., Chen, L., & Kahana, E. (2016). Mealtime interactions and life satisfaction among older adults in Shanghai. *Journal of Aging and Health*. https://doi.org/10.1177/0898264316641080.

Yi, C., & Huang, Y. (2014). Housing consumption and housing inequality in Chinese cities during the first decade of the twenty-first century. *Housing Studies, 29*(2), 291–311. https://doi.org/10.1080/02673037.2014.851179.

Ying, Z., Ning, L. D., & Xin, L. (2015). Relationship between built environment, physical activity, adiposity, and health in adults aged 46–80 in Shanghai, China. *Journal of Physical Activity and Health, 12*(4), 569–578. https://doi.org/10.1123/jpah.2013-0126.

Yun, J. (2015, June). *Shanghai Shikumen landscape: Weaving past, present, and future*. The 52nd World Congress of the International Federation, Petersburg, Russia. https://www.researchgate.net/publication/309486947_Shanghai_Shikumen_Landscape_Weaving_Past_Present_and_Future.

Zhang, F., Li, D., Ahrentzen, S., & Zhang, J. (2019). Assessing spatial disparities of accessibility to community-based service resources for Chinese older adults based on travel behavior: A city-wide study of Nanjing, China. *Habitat International*. https://doi.org/10.1016/j.habitatint.2019.05.003.

Zhang, J., & Weng, Q. (2006). Shikumen residences reveal Shanghai's past. *Chinese Female, 6*. Retrieved from http://en.cnki.com.cn/Article_en/CJFDTotal-ZFLW200606018.htm [Chinese].

CHAPTER 5

Fostering Community Caregiving Relationships

This chapter presents interaction and caregiving dynamics between elderly clients and service centers. We explored, how staff members in the service centers formed caregiving relationships with elderly clients through everyday interactions and service delivery. We interviewed 37 staff members from all the service centers in this HCBS program through in face-to-face, in-depth interviews to share their work experiences, their relationships with elderly clients, and their prospective views on the development of HCBS in Shanghai.

5.1 Formal Caregiving Relationships in the Community

Compared with residential care, most elderly clients prefer to live at home and in their familiar community, both in Western and Asian countries (Chen & Han, 2016; Thomas & Blanchard, 2009; Wiles, Leibing, Guberman, Reeve, & Allen, 2012; Wu, Carter, Goins, & Cheng, 2005). Although previous studies have suggested that the work experiences of HCBS care staff are crucial for the quality of care (Figueiredo, Barbosa, Cruz, Marques, & Sousa, 2013; Graham, Wing, Wadhams, & Worden, 2013), existing evidence has focused mainly on staff work experiences and their relationships with elderly clients in institutional settings (e.g., Brownie & Nancarrow, 2013). How staff members perceive the job of caring for elderly clients in the community has remained understudied.

L. Chen and M. Ye, *Community Eldercare Ecology in China*,
https://doi.org/10.1007/978-981-15-4960-1_5

Indeed, long-term care staff play an important role in the quality of life of elderly clients, especially through the various services they provide. It has been well documented that staff behaviors and attitudes in residential care settings are related to residents' quality of life, such as improving clients' physical health and cognitive abilities (Brownie & Nancarrow, 2013; Christenson, Buchanan, Houlihan, & Wanzek, 2011), decreasing depression and stress (Gallo et al., 2013; Melrose, 2013), helping pain management (Cimino, Lockman, Grant, & McPherson, 2014; Gagnon, Hadjistavropoulos, & Williams, 2013), reducing mortality rates (Potter et al., 1997), and providing care at the end of life (Goddard, Stewart, Thompson, & Hall, 2013; Munn et al., 2008).

Specifically, Gagnon and colleagues (2013) found that staff members placed importance on palliative care and symptom management for elderly clients who live in long-term care and assisted living facilities. Melrose (2013) pointed out that staff helped to improve elderly clients' well-being and reduce their stress after relocating to a nursing home. Potter and colleagues (1997) reported that staff members played a critical role in introducing vaccinations to elderly patients in medical long-term care facilities, which helped decrease mortality rates from 17 to 10%.

However, the majority of existing research has focused on staff members' work experiences in residential care or clinical settings (e.g., Christenson et al., 2011; Cimino et al., 2014; Goddard et al., 2013; Melrose, 2013; Potter et al., 1997; Reimer & Keller, 2009), while limited studies have examined community care settings. It should be recognized that working in a residential care setting is different from working in an HCBS program caring for community-dwelling elderly clients. Elderly clients living in nursing homes or other long-term care facilities usually have poorer health conditions than their community-dwelling counterparts, requiring a more structured life and more professional medical attention than caregiving (Curle & Keller, 2010; Graham et al., 2013; Thomas, 1996). Indeed, in a nursing home, elderly clients have to follow relatively strict daily routines, such as the time to eat or the location to walk. These elderly residents must comply with the facility's rules and regulations and staff members have power and authority over them (Thomas, 1996).

In contrast, HCBS staff do not have high levels of power over community-dwelling elderly clients, who often enjoy relatively good health and are able to address their own medical needs. Thus staff tend to respect elderly clients' autonomy and focus primarily on caring for and interacting with them. Considering the growing number of older adults who prefer to age in place rather than relocating to nursing homes worldwide (Thomas & Blanchard, 2009; Wiles et al., 2012), it is important to examine how HCBS staff care for elderly clients in various community settings.

A similar research gap is reflected in the fact that most studies related to work stress for long-term care staff also take place in residential care settings. These staff members face high levels of stress and heavy workloads, as well as high turnover rates (Figueiredo, Barbosa, Cruz, Marques, & Sousa, 2013; McGilton, Boscart, Brown, & Bowers, 2014; Westermann, Kozak, Harling, & Nienhaus, 2014). In a systematic review, Westermann et al. (2014) found that workplace stress and burnout were common for nursing staff members working in residential care facilities. Their review concluded that although financial incentives and professional interventions can help to reduce burnout among staff members, the effectiveness of these interventions is rather limited. As such, how work-related stress may influence these staff members' work performance and the quality of care they provide deserves further attention.

McGilton and her colleagues (2014) compiled a list of the main reasons staff members leave the long-term care field, including insufficient resources, unsupportive leadership, lack of professional development, and bad relationships with elderly residents (McGilton et al., 2014). When staff members had meaningful relationships with elderly residents, sufficient training, and promising professional promotion opportunities, they were likely to stay longer in the field (McGilton et al., 2014). Professional training and relevant interventions to improve caregiving skills, such as fostering dynamic staff–resident interactions, can help staff members feel empowered and supported (Ryvicker, 2011). For example, nurses' aides were aware that their caring attitudes could comfort residents (Bowers, Esmond, & Jacobson, 2000; Forsgren, Skott, Hartelius, & Saldert, 2016), while their own burnout could distress residents (Hunt, Corazzini, & Anderson, 2014). Intervention studies show that when staff members in the nursing homes learned story-sharing strategies (Heliker & Nguyen, 2010) and enhanced interpersonal skills (McGilton et al., 2003), their relationships with elderly residents improved significantly and their work-related stress decreased. Previous studies have also identified strategies that nurses' aides used to achieve positive caring relationships with older residents, including acknowledging residents' personal needs (Nakrem, Vinsnes, & Seim, 2011), getting to know residents (Brown-Wilson & Davies, 2009), and sharing personal information (Bergland & Kirkevold, 2005, 2008). These strategies can enhance the caregiving process as well as reduce prevalence of stress and depression for staff members in residential caring settings (McGilton et al., 2003).

However, little attention has been paid to staff members' working conditions and experiences in community long-term care settings. In fact, staff members in HCBS programs also face high levels of stress, burnout, high turnover, and other problems when caring for elderly clients on a daily basis (e.g., Howes, 2014; Rossi et al., 2012). In particular, Rickerson and colleagues (2005) found that staff members in community-based long-term care programs also suffered from a variety of depressive symptoms when elderly clients with whom they interacted passed away. Bereavement support was needed not only for elderly clients' family members but also for these staff members (Rickerson et al., 2005). Staff members' working experiences in community long-term care programs are likely informed by elderly clients' lives, well-being, and attitudes toward staff.

Because community-dwelling elderly clients usually enjoy relatively high functioning and independence, staff members in community service settings are able to have more time and freedom to interact with them, which in turn can facilitate better formal caregiving relationships. For example, community-based senior centers in the United States provide multiple services, such as counseling, meals, friendly visits, domestic assistance, and transportation to meet different needs among elderly clients (Gelfand, 2006). The majority of these services are provided directly by staff members from senior centers in the community (Gelfand, 2006). Wilson (2012) argued that the relationship between staff members and elderly clients is a continuum of person-centered care. Fostering such caring relationships moves beyond individual-level service provision, recognizing elderly clients' needs from a holistic perspective in the context of a community (Wilson, 2012). In addition, these staff members interact with elderly clients on a regular basis, which may offer some insight into the development of community-based eldercare.

Another research gap is the lack of literature on caring for elderly clients in culturally diverse community long-term care programs. Studies of staff members' work experiences in HCBS programs have largely examined Western contexts with European Caucasian elderly clients (e.g., Gelfand, 2006; Wacker & Roberto, 2013). In fact, both care receivers' and care providers' ethnic backgrounds and cultural norms can affect their caregiving expectations, interpersonal interactions, and the quality of care (Jones & van Amelsvoort Jones, 1986; Reimer & Keller, 2009). For this reason, staff members' work experiences in community long-term care settings should be analyzed in diverse contexts. At the same time, although studies have increasingly assessed the progress of long-term care in urban China (Chen & Han, 2016; Chu & Chi, 2008; Z. Feng, Liu, Guan, & Mor, 2012; Flaherty et al., 2007), limited studies have examined staff members' work experiences in Chinese HCBS programs. Chen and Ye (2013) investigated the relationship between elderly clients' motivation to receive community-based long-term care and their children's support in Shanghai. But this study primarily focused on evolving family caregiving in Shanghai, as opposed to social support in the community. Q. Feng et al. (2013) revealed that the increasingly available community-based eldercare in Shanghai might have contributed to trends in strengthened activities of daily living (ADLs) and instrumental activities of daily living (IADLs) among its aging population. But this study did not examine the implications of community-based eldercare.

Given the increasing size of the aging population in China, especially in Shanghai— the "oldest" city in China—and the shrinking number of adult children available for family caregiving (Chen & Ye, 2013; Cheung & Kwan, 2009), elderly clients' needs will become increasingly important and demanding (Andersen, 2008; Blair, 2012). This trend demonstrates the important role of HCBS staff, because they provide direct care for elderly clients on a daily basis in community long-term care settings. Their work experiences, perspectives, and feelings can connect elderly clients with the HCBS program on the meso level in the community ecological system. Their work experiences can also offer insight that can be used to improve the quality of care for elderly clients as well as the development of future service programming.

5.2 Staff Members' Work Experiences in the HCBS Program

5.2.1 Interview Process

We began interviews with staff members of the HCBS program in June 2014. First we explained the study purpose and procedures to the administrative board of the HCBS program. The head administrator gathered the staff deputies from each service center for a briefing session. We, the authors, attended the sessions to answer questions and address concerns. Then we visited each service center to interview staff members from July to September 2014.

The criteria for study participation included working in the program for at least six months and providing direct services to elderly clients on a daily basis. The rationale of choosing six months as a cutoff was that this period would likely provide a general idea of the overall service provision procedures for staff in the program. When we met any staff members who worked in the program for less than six months, we did not interview them, but we explained our visit and let them know about the study. We interviewed one staff member per service center. The final sample size was 37.

Among the 37 participants, three of them were men and 34 were women. Their average age was 50 years old, ranging from 28 to 62. On average they had worked for 25 months in the program, ranging from 10 months to eight years. Three participants had finished college, 33 had received their high school education, and one of them had finished elementary school only. Most staff members lived close to the service centers where they worked, within 10–30 minutes' walking or cycling distance. Three of them lived far away, taking them two hours by bus or subway.

We interviewed each staff member in a private room at the service center where he or she worked. Each interview lasted for one to two hours and was audiotaped with his or her permission. We opened the interview with a grand-tour question: "How do you like working in the program?" Then we went into detail about each service item they provided, their relationships and interactions with elderly clients, their collaboration with other organizations in the community, and their suggestions for improving the program. Sample questions included, "How do you think elderly clients like various activities offered in your service site?" "During the time of working here, how is your own relationship with these elderly clients?" and "If you were able to design an additional service item/activity for elderly clients coming to your service site, what would that be?" Although not originally on the interview guide, almost every participant raised the topic of work ethic. We asked each staff member to explore their devotion and motivation for their work. Interviews were conducted in Mandarin and Shanghai dialects.

5.2.2 *Providing Demanding Services*

In general, heavy workloads. Given limited government subsidies and the non-profit nature of the HCBS program, every service center was only able to employ two to three staff members based on the number of elderly clients signing up for services. Facing an increasing number of elderly clients' using HCBS services, staff members' workloads steadily accrued to the point of being unbearable. Almost all the staff members considered that they were working on the brink. For example, Ms. Wen (all staff members names are pseudonyms) said:

> My colleague and I can manage the work when both of us work in the service center. But if she takes leave, I can barely finish the work. Every morning, I tell myself I have to leave home early. My husband often laughs at me and says, "When don't you leave early?" There are so many things to cover every day. For example, today I need to work on a report, tomorrow I have to prepare meals. At the end of a month, we need to collect money from each elderly client to settle their meal plans for the next month. These days are the busiest days of the month. We have to collect money and write the financial report. I try to start early and finish late but there is always work.

Because of such demanding workloads, all staff members mentioned the high turnover rate among their newly hired colleagues. For example, Ms. Wu expressed stress about the situation:

> We are running out of workers. Younger people come and go instantly. They are afraid of the workload. They are not willing to talk to elderly clients. Before they came here, they had no clue what it was like working in community-based long-term care programs. Once they've worked here, it's too much for them. They cannot continue. They complained a lot and then left… I am not saying that young people are not good at their work. Some of them do not pay careful attention to this kind of work. They should be more attentive. I know everybody who comes here, their family members, both old and young, both living and passed away, their homes, their marital status, and their jobs. I am familiar with each of the elderly clients who come here. That is what I call working in eldercare.

Although some of them had worked in the program for a long time, all participating staff members expressed worry about the declining workforce and the sustainability of the program. Ms. Wu continued:

> [Working in HCBS programs] is actually very promising. We are going to have so many elderly clients in Shanghai. There will be no shortage of work opportunities. Those who left just did not see an immediate outcome. They needed quick money without much conscience. They should have known more about the security of this job and the importance of this career.

With the digitization of office work, most middle-aged staff members with fairly low education levels found it very challenging to use a computer. Their digital work included compiling weekly activity schedules, monthly meal plans, and occasional day trips for elderly clients. Because HCBS program does not offer computer training for them, staff members had to spend extra time learning computer skills after work. As Ms. Pan generalized,

> We have to have a body like a young athlete and master computer skills like an undergrad. I have to climb several 7-storey buildings without elevators to deliver lunch and dinner. We are in great need of younger people to do office work and deliver meals. We are also in great need of computer knowledge, which I hope someone can teach us.

The overall workloads remained both physically and intellectually demanding for staff members. Given these conditions and the high turnover rate among younger workers, staff also worried about whether there would be sufficient workforce to provide care for the increasing number of elderly clients, as well as about the future development of the HCBS program.

Meal service most popular. While all staff members agreed that meal service was the most important and popular service among elderly clients, this was the most demanding part of their daily work and required most of their attention. For example, a typical day for Ms. Wang began at 6:00 a.m. in the service center. About 15 elderly clients came for breakfast at 7 a.m. sharp. Ms. Wang and her colleague had to prepare their breakfasts in under an hour. At 10:30 a.m., nearly 100 elderly clients came to pick up lunch or eat lunch onsite. Given limited space in the service center, Ms. Wang and her colleague divided elderly clients into three groups of about 30 each to come at different times for dining onsite. More than 20 elderly clients usually stayed at the service center for activities after lunch. At 4:00 p.m. another group of 10 elderly clients came to pick up dinner. The number of meals that each service center provided varied from 50 to 200 every day and the number of elderly clients signing up for meal service had been growing steadily in recent years.

In addition to preparing food for dining onsite, staff members delivered meals to clients over 80 years old and those with disabilities, which they admitted was particularly challenging. For example, Ms. Shen described her strenuous delivery trips:

> Most elderly clients who need delivery in our service centers live on the sixth or seventh floor without elevators. I have to climb up and down. I ride a bike to deliver. It takes about an hour and a half for me to deliver meals to eight elderly clients when the traffic is good. When the traffic is busy, it may take up to two hours. This is too much for me. I am not young anymore. I am already 60. I also should be the one who can have elderly service here. I'm still providing services for others.

Because most of the staff members were over 50 years old, they felt that providing meal service for an increasing number of elderly clients onsite or by delivery was beyond their capacity. The meal plan arrangements, delivery to elderly clients' homes, and serving food onsite in the service centers posed physical and intellectual challenges for staff members. Furthermore, two-thirds of the 37 service centers cooked rice themselves and had other dishes delivered from a central kitchen. Ms. Ye was willing to take responsibility for such cooking:

> Freshly cooked rice can make our clients feel at home, you know, home-made meals. Most of our clients are empty-nesters. They long for a family atmosphere, which is our goal to provide in the service center. Although it is time-consuming and adds extra work, I would still prefer to prepare freshly cooked rice for elderly clients.

Staff members believed that the average workload of preparing, serving, and delivering meals sometimes exceeded their abilities, potentially draining them physically and mentally. At the same time, staff were still willing to invest time and effort to provide attentive meal service, to make elderly clients feel at home.

Ms. Ye's HCBS program in particular boasted about individualized meal plans and flexible meal schedules to meet elderly clients' convenience, at great costs to staff members in terms of effort and time. The program offered three different meal plans at 8 yuan, 12 yuan, and 20 yuan (i.e., about \$1.10, \$1.70, and \$2.90 USD). Elderly clients could combine different lunch and dinner plans. Staff members commented that they had to be extremely careful about these meal plans. For example, Ms. Sun was stressed over the various combinations of elderly clients' meal plans:

> Someone eats only on weekdays; someone eats only on weekends; someone wants both dinners and lunches; and someone only takes lunches; someone wants lunch on weekdays and dinners on weekends… you name it. We can get countless combinations from elderly clients and each month they can come up with new combinations, and some of them can just change their mind. Given the huge amount, it takes us at least 2-3 days to go through everything and check with each of [the elderly clients] to make sure the number is correct for each month. The beginning of each month is like a battlefield.

Nevertheless, during each month, elderly clients often changed their minds and wanted to switch to different meal plans after staff members had submitted the plans to the central kitchen of the HCBS program. Elderly clients were also able to change their meal plans two days in advance when they had other engagements or just changed their minds. Staff members would calculate the actual number of meals for those who changed their plans during the month and refund them the following month. Settling meal plans with each elderly client was complicated and time-consuming. Staff members made sure that every elderly client received the right meal plan and the right amount of refund, if any. They felt responsible for these clients. As Ms. Wu described,

> Not many elderly clients live with their children. The empty-nesters need our support. Elderly clients need to know that there are people who care about them. They are not left alone. They come here every day. [Staff members] are basically their family members. I have to protect them as long as I am doing this job. That's what keeps me doing this much work.

As such, despite such heavy workload, staff members tried to keep up with the increasing number of elderly clients who signed up for meal service in recent years. They viewed meals as a fundamental instrument for providing care and support. Homemade meals, in particular, created a family-like atmosphere for elderly clients to feel included and cared about. At the same time, staff members believed that well-prepared meals could also be helpful for elderly clients' health conditions. For example, Ms. Sun shared her observation:

> A number of elderly clients in our service center suffer from diabetes and some of them did not strictly follow doctors' prescriptions or adhere to the medicine. We have to watch out for them. The central kitchen [of this HCBS program] provides special meals tailored to diabetic patients. After having meals here, elderly clients' blood sugar became more stable than when they cooked themselves.

Indeed, as indicated in an old Chinese saying, "Food is living" (*Min yi shi wei tian*), nutritious meals are the cornerstone of elderly clients' health.

Nine other equally important services. Staff members reported to have diligently fulfilled all other nine services offered in the HCBS program. All service centers provided laundry facilities, bathing assistance, health promotion lectures, reading groups, knitting groups, dancing groups, as well as setting up space in the service centers for subcontracted services, including pedicure and haircut at least once a month. During this process, staff members were attentive to elderly clients' service preferences; they spontaneously attended to elderly clients' needs. They noticed that most elderly clients who over 80 years old seldom went outside. Ms. Zhao arranged day trips for these elderly clients twice a year to do sightseeing in the city. She recalled the first trip:

> It was always frowned upon to take these old-old clients on trips because they were fragile. No one had done that before. The HCBS program and all of us were afraid of taking responsibility. It is understandable that if someone falls, there will be a very large medical bill. When several elderly clients talked with me about their isolated life, I had this bold idea that those who were capable of walking around should have a chance to go out and see the city. Elderly clients should not be excluded from society, regardless of their advanced age. We did not go very far, just to the other side of the Bund in Pudong New District and the Oriental Pearl Radio & TV Tower. The 20 elderly clients laughed like babies. We got discounted tickets to the TV tower and had a great time up in the air.

Staff members also negotiated with subcontracted service providers to increase service frequency and provide services at elderly clients' homes for those who were disabled or fragile. For example, Ms. Wu was proud of the fact that she had successfully increased the frequency of pedicures for all elderly clients in her service center from once a month to once a week, which expanding coverage and increasing convenience for elderly clients. Ms. Zhao elaborated on the extensive list of activities arranged in the service center:

> In addition to meal service, we have a doctor come to measure blood pressure every Monday. On Tuesdays we have Shanghai Opera group—[elderly clients] sing quite beautifully. Wednesdays we usually offer health promotion lectures. The rest of the week, elderly clients have dancing groups, such as ballroom dance. They have activities planned out every day. A group of older gentlemen play bridge every day in the small room next door, even on the weekends. We see them every day like a family. If someone does not show up, we will call them or call their children to make sure they are safe and sound.

Similar to meal service, staff members were particularly keen to help elderly clients maintain their health conditions and functional levels. Each service center offered blood pressure measurement once a week by inviting physicians from the community clinics. The HCBS program collaborated with local hospitals to provide free physical exams once a year for elderly clients who were over 80 years old. Staff members also regularly visited elderly clients who lived alone or those who did not have children to closely monitor their health and safety. For example, Ms. Fang checked on about 20 clients without children every day:

> In the afternoon, I visited 20 clients who live alone, until evening. These were our long-time clients. I know them very well. We always talk and share our lives. There used to be an old man who moved away but kept coming back to our service center. He wanted to stay because of our service center and our care. That meant a lot to me personally. The value of our work is shown in these examples.

This HCBS program strived to create a family atmosphere for elderly clients. Although their workload continued to grow, staff members valued their work, demonstrated their genuine care for elderly clients, and shared their concerns about the sustainability and high turnover rate of community-based long-term care in Shanghai.

5.2.3 Growing Emotional Attachment

Building rapport with elderly clients. Despite physical, mental, and financial challenges, staff members emphasized sharing a special bond with elderly clients. All staff members, following this HCBS program's mission, were keen to create a family-like atmosphere for elderly clients and to treat them like family members. More importantly, staff members practiced caregiving similarly to the way they cared for their own elderly parents. By complying with filial piety, staff members viewed caregiving as a moral obligation as well as helping to carry on this crucial social and cultural tradition. For example, Ms. Yang described her relationship with elderly clients in her service center:

> Elderly clients always check in with me about wherever they are going. They like to share their whereabouts with me very much, you know, to give me a heads-up. This is like what I do at home with my own parents. When you leave for work, you say goodbye and when you come home, you say "I'm home." When some elderly clients go travelling with their children, they come to say goodbye to me. When they come back, they bring [home] souvenirs. Sometimes they even check in here when they just go to their children's over the weekend. Checking in with [the service center] brings a closeness between us. The way we provide care follows filial piety, which is very precious.

Staff members insisted on knowing what was going on with every elderly client in the service center, and their whereabouts. Their feelings of responsibility and duty prompted their efforts to strive for high-quality care.

Staff members also respected elderly clients' various characteristics and provided corresponding services. They sought extra financial support for elderly clients with low incomes, introduced personnel suitable for home care to clients with disabilities or low functioning levels, and encouraged clients who lived alone or did not have children to participate in more activities at the service centers. For example, Ms. Pan shared her thoughts on making this effort:

> It is important to know each of the [elderly clients] well to understand their actual needs. Elderly clients are not all the same. It is impossible to aggregate them into one group simply because of their ages or functional levels. That just oversimplifies everything. It is important to recognize the

> differences in their characteristics and personalities, in addition to each individual's physical, mental, and cognitive levels. Many elderly clients are quite idiosyncratic. Each of them is one-of-a-kind. Not until you get really close to them individually and get familiar with their family, can you say that you know them individually. Then, you may get even closer with them, to learn more about their needs. Working here really takes time.

With such continuous and genuine effort to improve service delivery and increase the quality of care in the HCBS program, staff members prioritized efforts to acknowledge elderly clients' actual needs and tried to meet all these needs.

This family-oriented service delivery style in the HCBS program further fostered mutual care between staff members and elderly clients. For example, Ms. Wu said,

> I went on a family trip for about a week last month and did not tell many [elderly clients]. After I came back, almost all of them asked me about the trip and some of them were so happy to welcome me back, "Where have you been? I miss you!" The attachment between us is real and strong.

Mutual care also helped staff members to gain elderly clients' trust. Clients shared their family issues with staff, mostly focusing on their relationships with their adult children, and sought staff members' suggestions. For example, Ms. Li shared one of the stories:

> A granny often fought with her daughters-in-law. Her daughters-in-law came to us for assistance to soothe their mother-in-law. They considered us all family members. We talked with the granny and tried to mediate among them. We made some progress along the way. I told the granny that the [daughters-in-law] work during the day and care for their own children during the night. They do not have much time to spend time with the granny. However, it doesn't mean that her daughters-in-law do not care about her. The older generations should also be reciprocal and recognize what children are doing. Mutual understanding is important. We also apply mutual understanding in our work and elderly clients appreciate us.

Staff members identified the organic rapport and mutual understanding with elderly clients as a meaningful reward for their diligent work.

Mutual affection and care further encouraged elderly clients to develop a sense of belonging to the HCBS program and the community, fostering their growing involvement in the community and expanding their social networks. For example, Ms. Ye noticed a growing number of elderly volunteers in the community around her service center:

> Many of our clients who are still physically capable became more and more involved in the community. They participate in many activities, such as neighborhood watch and residents' organizations. They see themselves as still able and valuable. That is very important to them. I think the usefulness of elderly clients is critical for both their physical and mental health. Elderly clients need to know that they are still worthy and appreciated. The service centers [of this HCBS program] cultivate this understanding and encourage all of our clients to stay engaged with us and with the whole community.

Building rapport between staff members and clients and increasing attachment between elderly clients and their community not only enhanced the quality of care of the HCBS program, but also expanded elderly clients' social networks and increased their sense of value, which strengthened their social inclusion and ultimately, their quality of life.

Prioritizing elderly clients' best interests. By knowing elderly clients' needs and taking their perspective, staff members tried to meet their needs and provide corresponding services as much as possible. Many staff members reported that they had provided extra services for elderly clients when they noticed that they were in need, regardless of whether or not the service was included in the HCBS program. For example, given the current popularity of e-commerce in China, several service centers have begun to offer online shopping since 2013. Ms. Zhu described how it worked:

> Some elderly clients know how to browse websites and they take notes about what they want. They give us these notes and we use our Apple Pay or AliPay to help them to purchase online and have things delivered to the service center. Elderly clients come and pick it up or we deliver to them when things are heavy, and then reimburse us. Although this service is not listed in our job description, it is very convenient and intuitive. Elderly clients should not be left out of society. We encourage more elderly clients to do the same and expand their understanding of what is going on currently. It does them good, you know, to stay connected with society. In our service center, it works out perfectly.

Despite the extra work, staff did not consider elderly clients' needs to be limited to the 10 types of service the HCBS program offered.

Despite their heavy workloads, staff members encouraged more elderly clients in the community to participate in the program. For example, Ms. Ye said,

> I keep telling elderly clients to introduce their friends in the neighborhood who have not yet used our program. [Community-dwelling elderly clients] should come and visit. We try to cover all aspects of their lives, including dining, activities, socializing, health, and so forth. Elderly clients can benefit so much that they should take advantage of the community-based long-term care.

When they acknowledged the benefits of the HCBS programs, staff members were motivated to recruit more elderly clients. They believed more community-dwelling elderly clients should have HCBS support. Staff members also proactively reached out to resident neighborhood associations (i.e., semi-governmental organizations that handle day-to-day civil affairs issues in the community) to organize more activities in the service centers, such as old movies and ping-pong. In doing so, staff members considered their work more like a mission than an obligation, were motivated to hone their caregiving skills, and prioritized the best interests of their elderly clients.

Encountering difficulties. Despite their enthusiasm, staff members still expressed downsides to their work. One of the most significant difficulties in providing services was that some elderly clients took their work for granted. Some began to rely heavily on staff members, instead of their adult children. Staff members felt this situation was unfair because HCBS was supposed to remain supplementary to family caregiving, not replace it. For example, Ms. Cheng recalled one incidence:

> One old lady once said, "My son is too busy to take me to a doctor's appointment. You should take me to see the doctor." I was astonished. I was not her daughter. I cannot just step into fill her children's position in addition to my full workload at the service center. Elderly clients should stay active and try not to rely on us for everything, especially medical issues. We cannot assume such huge responsibilities. This is beyond our job descriptions and responsibilities. We are here to assist, not to replace.

Staff members were frustrated about such unappreciative attitudes from elderly clients. They considered their primary job as assisting elderly clients with their daily routines, but not their entire lives. Staff members insisted on the supplementary role of HCBS and predominant role of family members when providing care for elderly clients. They believed that the Chinese family caregiving tradition should still be valued and that adult children should take responsibility as primary caregivers for their elderly parents, rather than taking advantage of community-based long-term care with limited staffing and resources. Ms. Pan called this kind of behavior "community abandonment":

> Adult children sort of dump their elderly parents with us. What for? Their parents are still theirs, not ours. We try our best to take care of these elderly clients in the community outside their homes with the hope of improving their general well-being. It is wrong to forget about our Chinese family values. The older generation is still an asset for society. Their children should know better than this.

Another difficulty in providing services was that some elderly clients were difficult to deal with. For example, Ms. Wang said, "Elderly clients get confused easily and they are stubborn. I'd be lying if I said there were no arguments." Uncooperative elderly clients existed in every service center. All staff members had dealt with them using various interactive strategies, such as negotiating, storytelling, and confession. They tried to have these elderly clients be on the same page with them. Some staff members noticed that these clients usually showed signs of memory loss and disorientation. At the same time, staff members admitted that memory loss and personality changes were part of the aging process. They strived for accommodating for these elderly clients and avoided friction as much as possible, though it took time to know each elderly client's personalities and preferences.

In addition to challenges in caring for elderly clients, staff members complained about low salaries. All staff members felt that low financial incentives did not compare with their hard work. They were paid minimum wage in Shanghai, 1580 RMB (i.e., about $225) per month in 2018, including health insurance, but without any overtime payment. For example, Ms. Cheng pointed out:

> Our salary is unable to match up with what we do every day, not even close. I do this job based on my conscience. I love this job but I am not doing charity. We deserve better pay and wider recognition. The number of elderly clients is growing so fast in Shanghai that the government should consider how to attract more people to this field. With the current salary, it will likely just scare people off.

Indeed, given the situation in the service centers, all staff members usually worked overtime, because elderly clients' needs were not limited to working hours. Receiving inadequate compensation for their excessive workloads left staff feeling devalued.

5.2.4 *Working for Tomorrow's Older Self*

Despite heavy workloads, low salaries, and frustration with some elderly clients' unappreciative views on their service, staff members were proud of their work, which contributed to elderly clients, to the HCBS program, and to the community. They were motivated to provide better care for elderly clients, as Ms. Mao pointed out:

> The important thing is to know what these elderly clients really need and provide corresponding, tailored care for them. That is why we have worked ourselves to the bone to learn a lot of skills and tried to get trained properly. We are going to get old as well and I am now 61 years old. I will need the same kind of HCBS services eventually.

Staff members believed they could contribute to improving the quality of care of HCBS for the increasing number of elderly clients in Shanghai. They were a critical part of the growing popularity of HCBS programs. They were determined to enhance the quality of care and raise the service standards for future community-based eldercare in Shanghai because they were likely going to need this kind of services in a decade or two.

Staff members' views on HCBS were rooted in their perspective on their own future: They worked for their future older selves and wished to receive high-quality care when they, too, became old. For example, Ms. Dong said, "Elderly clients always say that God knows what we are doing. We will be repaid for what we are doing now." Staff members believed that their good deeds would be recognized by others and their work performance would be repaid by improved quality of care for future elderly clients—themselves. Some staff members also thought that the reason current elderly clients enjoy the good services in the HCBS program was that they had done something good to deserve this in their old age, what Ms. Zhou described as "a lucky aging cohort":

> These elderly clients went through so much turmoil before 1949. They worked hard to build our country. Now, they deserve a good later life. We are here to help, you know, one good turn deserves another.

Another factor related to their devotion was that most staff members were mothers of an only child. Given the limited number of children who can support family caregiving, staff worried about their own future long-term care. For example, Ms. Wang said:

> [Because] our children are all only children, I am not sure how they would manage when we are old. Children will have their own family and life. We, as old adults, would be in the way. We should find alternatives to solely relying on children.

Staff members foresaw the rising demand for community-based eldercare in Shanghai. It made sense for them to consider an alternative to family caregiving and seek additional caregiving options.

Based on their long-term work in the HCBS program and their own expectations for future community-based eldercare, staff members were keen on offering sensible suggestions for service improvement. First, they advocated higher quality of care for elderly clients. For example, Ms. Zhang said:

> The number of elderly clients that I provide services for is not important. What matters most is the quality of care. We need to be more attentive to elderly clients' needs. Elderly clients' health conditions can change so quickly that we should stay cautious and alert. Only by always keeping a close eye on them shall we improve the quality of care.

Second, in addition to resident neighborhood associations, staff members wanted to extend service collaboration with more volunteers and service providers in the community, such as personal care and home care workers. They thought it was worthwhile to explore as many community resources as possible for elderly clients. For example, Ms. Shen suggested,

> More collaboration with elementary and middle schools can benefit our elderly clients. It is a good opportunity for them to keep updated with society and the younger generation. Children can also benefit from the wisdom and experiences of elderly clients. It's a win-win.

Third, staff members advocated elderly clients' empowerment. For example, Ms. Li suggested, based on her observation,

> Elderly clients should have the opportunity for designing and organizing some activities in which they want to participate. In doing so, we can learn what they want and whether the way we are doing it is right.

Finally, staff members suggested expanding the HCBS program to establish more service centers, to face the increasing number of elderly clients and growing use of elderly clients' services. They noted that the HCBS program was a valuable model for all the elderly clients in Shanghai. For example, Ms. Gu described her vision of HCBS programs in Shanghai:

> My colleague and I try to compile our experiences and insights into a service manual. We also take notes on every elderly client—their likes, dislikes, food allergies, hobbies, and health conditions, their medication, and so forth. All this information helps us to gather and practice useful strategies in the service center. Then all the service centers exchange manuals to hone our skills. We also connect with resident neighborhood associations to build up an HCBS network for elderly clients in the community. In doing so, the entire community can provide a caring environment for all the elderly clients living here.

Thus, because of their continuous involvement in the HCBS program and community-based long-term care services, these staff members recognized the potentially increasing demands for social support among community-dwelling elderly clients. They not only advocated the importance of developing HCBS programs for elderly clients in Shanghai, but also participated in strengthening the quality of care and shared the vision of developing such HCBS programs. Their effort was for their current elderly clients as well as for their own older selves.

5.3 Reciprocal Caregiving Relationships in the HCBS

Staff members and elderly clients in the HCBS program shared reciprocal caring relationships, consisting of two closely connected elements (Fig. 5.1). The first element related to staff members' daily work in the service centers. They administered care to the rising number of elderly clients seeking support from the HCBS program. For example, staff members endeavored to make meal service individualized and flexible for each elderly client. They proactively attended to clients' various needs, such as measuring blood pressure and organizing group activities. Staff were likewise attentive to elderly clients' daily lives and were willing to go beyond their job descriptions to help them. For example, they helped elderly clients shop online and arranged day trips for physically capable clients.

Staff members' diligence and attentiveness fostered strong emotional bonds with their elderly clients. This family-oriented approach facilitated frequent and intimate interactions between these two generations, establishing mutual affection and care. Frequent interactions with elderly clients familiarized staff with their needs and enabled them to provide responsive service. As a result, elderly clients developed a strong trust in

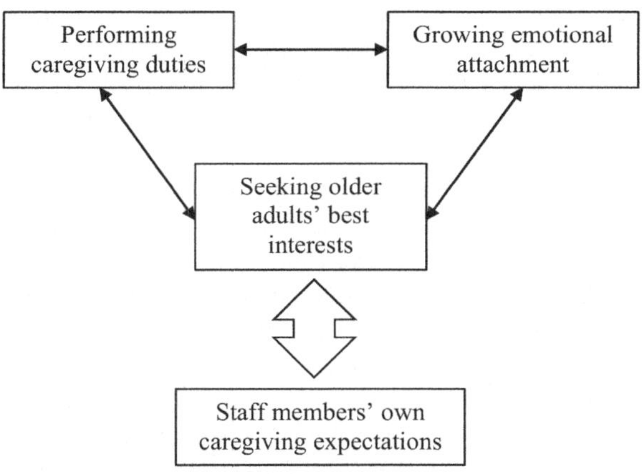

Fig. 5.1 A reciprocal caring relationship

staff members, sharing their feelings and family issues. This trust, staff members believed, symbolized their success in caring for elderly clients.

Reciprocating elderly clients' praise and trust, staff members' work ethic strengthened. Although staff sometimes felt frustrated about their daily custodial work and low salaries, they found their work meaningful and honorable because they cared about the older generation's well-being and the importance of HCBS. Prioritizing elderly clients' best interests became a part of their work ethic.

The second element of the reciprocal caring relationships related to staff members' own caregiving expectations. Their experience working with older clients informed their conception of quality care. They not only highly valued their work and caring for elderly clients but wanted to enhance the quality of eldercare they would someday receive. In the context of Chinese culture, staff members considered their work good deeds that would be repaid when they got older. This belief stemmed from the notion of karma, an important concept in Chinese culture that originates in Buddhism (Zhang, 2007). According to karma, an individual's fate is based on his or her previous deeds in the past life (i.e., reincarnation) or in earlier years of their current life (Zhang, 2007). The determinants of individuals' good fortune are how many good deeds they have completed. In other words, "one good turn deserves another" (*Hao ren you hao bao*). The good or bad results an individual experiences are determined by their prior good or bad deeds (Koo, Tin, Koo, & Lee, 2006). Consequently, staff members believed that their good deeds (e.g. providing quality service) could bring them improved HCBS support when it was their turn to receive care, and were motivated to improve the HCBS program's quality of care.

To further understand the relationships between HCBS staff and their elderly clients in our analysis, we applied a dyadic perspective (Kahana & Young, 1990; Kenny, Kashy, Cook, & Simpson, 2006; Rusbult & Van Lange, 2008). A dyadic perspective suggests that both entities in a dyad affect one another (Kenny et al., 2006). In a dyad, when both entities share a positive emotion during communication and interaction, an emotional connection may take place between them (Kenny et al., 2006). Indeed, a dyadic perspective on interactions between care providers and care recipients should be at the center of caregiving studies (Kahana & Young, 1990).

First, interpersonal interactions between staff members and elderly clients created a foundation for mutual respect. Frequent interactions between them helped to construct caring, stable, effective, and reciprocal relationships (Rusbult & Van Lange, 2008). Emotional reciprocity and support can improve elderly clients' psychological well-being in addition to the instrumental support they receive (Kahana & Young, 1990).

Second, interpersonal interactions between staff members and elderly clients facilitated service provision. Frequent and friendly interactions with elderly clients became an important channel for staff members to familiarize themselves with these clients and provide care that corresponded to their needs. At the same time, elderly clients' reactions to services provided feedback that helped improved staff members' service provision. When elderly clients showed positive reactions, staff members knew they could carry on the service. When they expressed concerns about a certain service, staff members were able to adjust in accordance with their clients' preferences. In particular, continuous communication with clients helped staff members know what clients liked or disliked—pointing especially their unmet and under-met needs. As such, interpersonal interactions became a foundation for better-targeted service provision.

Furthermore, staff members' understanding of the HCBS program's family-oriented service approach echoed filial obligation in Chinese family caregiving. Although staff members were not elderly clients' relatives or children, they were determined to take good care of them. Staff members' strong motivation may also have been the result of elderly clients' positive feedback and the increasing popularity of the HCBS program among elderly clients. In this caring, interdependent relationship, staff behaviors and outcomes may have been influenced and predicted by elderly clients' actions (Kenny et al., 2006). This dyadic perspective analysis suggests that the caring relationships with elderly clients remained at the center of their work experiences.

Staff members were motivated by their future aging selves, who would also need caregiving someday. Although they faced high expectations from elderly clients, they were motivated to change the HCBS program and its related communities: to make Shanghai an age-friendly city. They were proactive and worked on their own agenda; they did not wait for the city to become age-friendly. They believed that they had the power to shape communities by providing a family-oriented atmosphere and seeking the best interests of their elderly clients.

Work motivation theory offers a perspective through which staff members' strong motivation can be interpreted. This theory suggests that workers in public sectors do not work just for money or for fun (Kaufmann, Schulze, & Veit, 2011; Sussmann & Vecchio, 1982). According to a study by Kaufmann et al. (2011), a worker's motivation can be categorized into intrinsic and extrinsic motivations. Intrinsic motivations are enjoyment-based (e.g., learning a variety of skills on the job) and community-based (e.g., community identification). Extrinsic motivations involve immediate payoffs (e.g., payment), delayed payoffs (e.g., human capital advancement, such as offering training for staff members), and social motivations (e.g., external social values, such as the meaning of the work).

In our study, although the staff members did not have high salaries, they had external work motivations, including acting in accordance with external social values. Filial piety, for example, dictates that the younger generation take care of the older generation instrumentally, financially, and emotionally, regardless of any costs and under any circumstances (Chen, 2016; Cheung & Kwan, 2009; Chou, 2011). This concept has been considered the most important virtue of Confucian culture (Fei, Chang, & Ward, 1946) and has regulated Chinese society in terms of parent-centered caregiving and social norms (Ikels, 2004). Because staff members believed that treating their elderly clients like their own parents would enhance the quality of their caregiving, filial piety played a critical role in creating a family-oriented atmosphere for service centers and cultivating caregiving relationships in the HCBS program. Although staff members were not elderly clients' relatives or children, they were determined to take good care of them.

The emphasis on filial piety among staff members could be related to the fact that they were parents of only one child. The one-child policy, launched in the 1980s (W. Feng, Gu, & Cai, 2016), has challenged the parent-centered caregiving tradition regulated by filial piety. As a result of the One-Child Policy, 300 million families, accounting for 70% of households in China, had only one child between 1979 and 2015 (National Health and Family Planning Commission of the People's Republic of China, 2014). Meanwhile, Chinese legislation requires every adult child to take care of their older parents (Standing Committee of the National People's Congress, 2019). Many elderly parents now worry about how their only child will take care of them with no brothers or sisters on whom they can rely (W. Feng et al., 2016). Because of their experiences with the HCBS program, staff members believed in community-based care, as opposed to being afraid of relying on an overburdened only child. This motivation drove them to enhance the quality of services for elderly clients because they believed that their work could improve the service standards of the overall long-term care field. They looked forward to enjoying similar or even better services from the HCBS program when they retired. Thus, supplying high-quality service to elderly clients was tantamount to investing in their own future welfare.

5.4 Understanding Meso- and Micro-Level Interactions in the Community

The interactions between service centers' staff members and elderly clients suggest two levels of interactions in the social ecology of HCBS, meso and micro levels. All these interactions shaped these caregiving relationships.

On the meso level, HCBS regulations and resources affected staff members' working experiences and their relationships with elderly clients. Staff members faced heavy workloads with low compensation and insufficient resources. According to the institutional regulations of the HCBS in our study, all service centers need to provide at least 10 standard services for elderly clients, and the capacity to provide these services depends heavily on the ratio of staff members to elderly clients. However, each service center had only two staff members to provide all of the services. At the time of our staff interviews, there were 37 HCBS service centers in the Jing'an District. As such, only 74 staff members provided services for all the elderly clients in the district. While not every elderly client needed assistance from staff members, the workload was still extremely arduous, which made it difficult for staff members to maintain a high quality of service for elderly clients. Sreeramareddy and Sathyanarayana (2013) have pointed out that frontline health workers' heavy workload can lead to burnout and morale problems. Previous studies have demonstrated that burnout can cause low work satisfaction, absence from work, and a high turnover rate (Abrahamson, Jill Suitor, & Pillemer, 2009; Westermann et al., 2014).

Staff members reported that their work was both mentally stressful and physically taxing. In addition to their regular service work, they also performed other tasks, such as maintaining the service centers or writing reports for higher officers. Under the pressure of their heavy workloads, all staff members worked overtime. Most found little time to take a short break during the day. Worse, elderly clients often blamed staff for failing to deliver services according to their expectations, creating further emotional stress. Most staff members were concerned about conflicts between elderly clients' expectations and unmatched quality of care. These conflicts arose primarily from differences in expectations between elderly clients and staff members about how the work should be performed. Staff members said that many elderly clients had high expectations of the services the HCBS program offered. The positive correlation between high expectations and high stress among caregivers has long been established (Pratt, Schnell, Wright, & Cleland, 1985; Yuwanich, Sandmark, & Akhavan, 2016).

On the micro level of the HCBS social ecology, elderly clients' service satisfaction was foremost for staff members. Understanding care recipients' life experiences helps contextualize patients' expectations of care as well as manage their needs (Price, 2017). Our interviews revealed that staff members used their knowledge of elderly clients' life experiences to better communicate in their caregiving relationships; once they understood these life experiences, they were able to provide services tailored to individual clients. Each communication became an opportunity to increase the breadth of the staff member's knowledge. These personalized interactions bridged the gap between the HCBS program and elderly clients, building trust with the clients.

Staff members also provided customized services based on clients' characteristics, preferences, and health status. It is well established internationally that customized services offer a higher quality of care. For example, Swedish agencies recently launched a program to supply customized services in communities in which patients created a care plan alongside their family caregivers and health professionals to best meet their needs (Ekman et al., 2015). The theoretical basis of this program held that perspectives from health professionals, patients, family members, and caregivers are all equally important for the patient's well-being (Ekman et al., 2015). Evaluations of this program showed reduced anxiety and uncertainty in patients (Ekman et al., 2015). In our study, when receiving personalized services, older clients felt that staff members cared about them and enjoyed their relationship. These clients relaxed and were less anxious when interacting with staff members.

However, because they did not fully understand the long-term care system, some elderly clients believed that the HCBS program should provide whatever needs they had. When they were unable to receive sufficient support from their relatives, they pressured staff to meet their needs. True, elderly clients may have lacked knowledge about long-term care or the operation of an HCBS program, including what services a long-term care facility can or cannot provide. Many held expectations of staff members and the HCBS program that staff members could not reasonably satisfy. In reality, people often have high hopes for many public services without understanding those services (Yamamoto & Holloway, 2010). For example, staff members noticed that a few elderly clients simply requested whatever they wanted because an HCBS, as the name suggests, is an organization that provides services for elderly clients.

Such frustration discouraged staff members. They realized the stigma around the custodial work they provided yet also expected to receive respect and understanding. Public respect for frontline staff members is positively associated with their occupational satisfaction (Wadensten & Carlsson, 2001). Staff members in our study advocated for more understanding, more tolerance, and more care from society and elderly clients. They believed their salaries should reflect their hard work and their contributions to society.

Heavy workloads were also potentially obstacles to building relationships with elderly clients, given the time it takes to cultivate such relationships. Because of the limited amount of time devoted to an individual contact, staff members found it difficult to provide fully responsive services. When elderly clients asked for extra help, it was hard for staff members to meet all their demands. This finding is consistent with existing evidence that time pressure can negatively influence staff members' work (Gandoy-Crego, Clemente, Mayán-Santos, & Espinosa, 2009). Indeed, time pressure in the context of "confrontation with suffering, dealing with people in need of long-term care" (Westermann et al., 2014, p. 62), can further contribute to burnout for frontline healthcare workers.

In the long run, such heavy workloads can lead to negative outcomes for staff members' physical and mental well-being (Gandoy-Crego et al., 2009; Skirrow & Hatton, 2007; Sreeramareddy & Sathyanarayana, 2013; Westermann et al., 2014). A systematic review of occupational stress management interventions for nurses found that mindfulness-based stress reduction program and psychological training on mental illness can significantly reduce stress and burnout (Nowrouzi et al., 2015).

In sum, the meso- and micro-level environments as well as staff members' motivation appear to have had mixed impacts on HCBS services. At the meso level, staff members had to perform arduous work with low compensation and little caregiving support. These challenges negatively affected the quality of the HCBS services. At the micro level, dyadic interactions with their elderly clients may have improved staff members' quality of work when they received positive feedback. However, staff members also received negative feedback, which was challenging to constructing close caregiving relationships.

5.5 Chapter Summary

This chapter analyzed data from semi-structured, in-depth interviews with 37 staff members of the HCBS program. The interviews explored staff work experiences and their formal caregiving relationships with elderly clients. The chapter explicated the mutually dynamic interactions among staff members on the meso and micro levels of the social ecology of the HCBS program that shape, promote, or hinder staff members' ability to care for elderly clients. Based on staff members' descriptions, we offered a holistic interpretation of their daily work in the service centers. In particular, we highlighted the reciprocal relationships between staff members and elderly clients in the HCBS program.

We examined the work ethic that these staff members espoused in helping their elderly clients in the HCBS program. Despite a similar high turnover rate in Western long-term care settings, staff members in the HCBS program show a culturally specific Asian care protocol, rooted in filial piety, for their elderly clients. We also provided a glimpse of how these middle-aged staff members foresaw their own HCBS support in the near future. Given their interactions with current elderly clients, these future elderly clients offered a unique perspective on the development of HCBS in urban China. Finally, this chapter summarized how caregiving relationships formed in the HCBS programs can not only benefit the rapidly increasing community-dwelling elderly clients but also expand the overall reception of such services in urban China.

References

Abrahamson, K., Jill Suitor, J., & Pillemer, K. (2009). Conflict between nursing home staff and residents' families: Does it increase burnout? *Journal of Aging and Health, 21*(6), 895–912.

Andersen, R. M. (2008). National health surveys and the behavioral model of health services use. *Medical Care, 46*(7), 647–653. https://doi.org/10.1097/MLR.0b013e31817a835d.

Bergland, A., & Kirkevold, M. (2005). Resident-caregiver relationships and thriving among nursing home residents. *Research in Nursing & Health, 28*(5), 365–375.

Bergland, A., & Kirkevold, M. (2008). The significance of peer relationships to thriving in nursing homes. *Journal of Clinical Nursing, 17*(10), 1295–1302. https://doi.org/10.1111/j.1365-2702.2007.02069.x.

Blair, T. R. W. (2012). "Community ambassadors" for South Asian elder immigrants: Late-life acculturation and the roles of community health workers. *Social Science and Medicine, 75*, 1769–1777. https://doi.org/10.1016/j.socscimed.2012.07.018.

Bowers, B. J., Esmond, E., & Jacobson, N. (2000). The relationship between staffing and quality in long-term care facilities: Exploring the views of nurse aides. *Journal of Nursing Care Quality, 14*(4), 55–75.

Brownie, S., & Nancarrow, S. (2013). Effects of person-centered care on residents and staff in aged-care facilities: A systematic review. *Clinical Interventions in Aging, 8*, 1–10. https://doi.org/10.2147/CIA.S38589.

Brown-Wilson, C. B., & Davies, S. (2009). Developing relationships in long term care environments: The contribution of staff. *Journal of Clinical Nursing, 18*(12), 1746–1755.

Chen, L. (2016). *Evolving caregiving in contemporary China: Two generations, one decision*. New York, NY: Springer.

Chen, L., & Han, W.-J. (2016). Shanghai: Front-runner of community-based eldercare in China. *Journal of Aging & Social Policy, 28*(4), 292–307. https://doi.org/10.1080/08959420.2016.1151310.

Chen, L., & Ye, M. (2013). The role of children's support in elders' decisions to live in a yanglaoyuan (residential long-term care). *Journal of Cross-Cultural Gerontology, 28*(1), 75–87. https://doi.org/10.1007/s10823-012-9185-y.

Cheung, C. K., & Kwan, A. Y. H. (2009). The erosion of filial piety by modernization in Chinese cities. *Ageing & Society, 29*(2), 179–198.

Chou, R. J.-A. (2011). Filial piety by contract? The emergence, implementation, and implications of the "Family Support Agreement" in China. *The Gerontologist, 51*(1), 3–16.

Christenson, A. M., Buchanan, J. A., Houlihan, D., & Wanzek, M. (2011). Command use and compliance in staff communication with elderly residents of long-term care facilities. *Behavior Therapy, 42*(1), 47–58. https://doi.org/10.1016/j.beth.2010.07.001.

Chu, L.-W., & Chi, I. (2008). Nursing homes in China. *Journal of the American Medical Directors Association, 9*(4), 237–243. https://doi.org/10.1016/j.jamda.2008.01.008.

Cimino, N. M., Lockman, K., Grant, M., & McPherson, M. L. (2014). Knowledge, skills, and attitudes in caring for elderly clients with advanced illness among staff members of long-term care and assisted living facilities: An educational needs assessment. *American Journal of Hospice & Palliative Medicine, 33*(4), 327–334. https://doi.org/10.1177/1049909114561996.

Curle, L., & Keller, H. (2010). Resident interactions at mealtime: An exploratory study. *European Journal of Ageing, 7,* 189–200. https://doi.org/10.1007/s10433-010-0156-2.

Ekman, I., Hedman, H., Swedberg, K., & Wallengren, C. (2015). Commentary: Swedish initiative on person centred care. *BMJ, 350*(feb10 14), h160–h160. https://doi.org/10.1136/bmj.h160.

Fei, H. T. U., Chang, C. I., & Ward, R. F. (1946). *Earthbound China: A study of rural economy in Yunnan.* London, UK: Routeledge.

Feng, Q., Zhen, Z., Gu, D., Wu, B., Duncan, P. W., & Purser, J. L. (2013). Trends in ADL and IADL disability in community-dwelling elderly clients in Shanghai, China, 1998–2008. *The Journals of Gerontology Series B: Psychological Sciences and Social Sciences, 68*(3), 476–485. https://doi.org/10.1093/geronb/gbt012.

Feng, W., Gu, B., & Cai, Y. (2016). The end of China's one-child policy. *Studies in Family Planning, 47*(1), 83–86. https://doi.org/10.1111/j.1728-4465.2016.00052.x.

Feng, Z., Liu, C., Guan, X., & Mor, V. (2012). China's rapidly aging population creates policy challenges in shaping a viable long-term care system. *Health Affairs, 31*(12), 2764–2773. https://doi.org/10.1377/hlthaff.2012.0535.

Figueiredo, D., Barbosa, A., Cruz, J., Marques, A., & Sousa, L. (2013). Empowering staff in dementia long-term care: Towards a more supportive approach to interventions. *Educational Gerontology, 39*(6), 413–427. https://doi.org/10.1080/03601277.2012.701105.

Flaherty, J. H., Liu, M. L., Ding, L., Dong, B. R., Ding, Q. F., Li, X., & Xiao, S. F. (2007). China: The aging giant. *Journal of the American Geriatrics Society, 55*, 1295–1300. https://doi.org/10.1111/j.1532-5415.2007.01273.x.

Forsgren, E., Skott, C., Hartelius, L., & Saldert, C. (2016). Communicative barriers and resources in nursing homes from the enrolled nurses' perspective: A qualitative interview study. *International Journal of Nursing Studies, 54,* 112–121. https://doi.org/10.1016/j.ijnurstu.2015.05.006.

Gagnon, M. M., Hadjistavropoulos, T., & Williams, J. (2013). Development and mixed-methods evaluation of a pain assessment video training program for long-term care staff. *Pain Research & Management: The Journal of the Canadian Pain Society, 18*(6), 307–312. https://doi.org/10.1155/2013/659320.

Gallo, J. J., Morales, K. H., Bogner, H. R., Raue, P. J., Zee, J., Bruce, M. L., & Reynolds, C. F. (2013). Long term effect of depression care management on mortality in elderly clients: Follow-up of cluster randomized clinical trial in primary care. *British Medical Journal, 346*, Article f2570. https://doi.org/10.1136/bmj.f2570.

Gandoy-Crego, M., Clemente, M., Mayán-Santos, J. M., & Espinosa, P. (2009). Personal determinants of burnout in nursing staff at geriatric centers. *Archives of Gerontology and Geriatrics, 48*(2), 246–249. https://doi.org/10.1016/j.archger.2008.01.016.

Gelfand, D. E. (2006). *The aging network: Program and services* (6th ed.). New York, NY: Springer.

Goddard, C., Stewart, F., Thompson, G., & Hall, S. (2013). Providing end-of-life care in care homes for older people a qualitative study of the views of care home staff and community nurses. *Journal of Applied Gerontology, 32*(1), 76–95. https://doi.org/10.1177/0733464811405047.

Graham, J., Wing, S., Wadhams, K., & Worden, M. (2013). *Staff perspectives and perceptions of the culture change movement: A closer look at person centered care and the Eden alternative in long term care facilities*. Master's Level Graduate Research Conference. Retrieved from http://digitalcommons.brockport.edu/.

Heliker, D., & Nguyen, H. T. (2010). Story sharing: Enhancing nurse aide-resident relationships in long-term care. *Research in Gerontological Nursing, 3*(4), 240–252.

Howes, C. (2014). *Raising wages for home care workers: Paths and impediments*. In the U.S. Department of Labor and Institute for Research on Labor and Employment at the University of California, Berkeley, A paper series commemorating the 75th anniversary of the Fair Labor Standards Act. Retrieved from: http://www.dol.gov/asp/evaluation/reports/FLSAPaperSeries.pdf.

Hunt, S. R., Corazzini, K., & Anderson, R. A. (2014). Top nurse-management staffing collapse and care quality in nursing homes. *Journal of Applied Gerontology, 33*(1), 51–74.

Ikels, C. (Ed.). (2004). *Filial piety: Practice and discourse in contemporary East Asia*. Stanford, CA: Stanford University Press.

Jones, D. C., & van Amelsvoort Jones, G. M. (1986). Communication patterns between nursing staff and the ethnic elderly in a long-term care facility. *Journal of Advanced Nursing, 11*(3), 265–272.

Kahana, E., & Young, R. (1990). Clarifying the caregiving paradigm: Challenges for the future. In D. E. Blegel & A. Blum (Eds.), *Aging and caregiving: Theory, research and practice* (pp. 76–97). Thousand Oaks, CA: Sage.

Kaufmann, N., Schulze, T., & Veit, D. (2011, April 7). More than fun and money. Worker motivation in crowdsourcing: A study on mechanical turk. In *Americas Conference on Information Systems (AMCIS)*. Detroit, Michigan.

Kenny, D. A., Kashy, D. A., Cook, W. L., & Simpson, J. A. (2006). *Dyadic data analysis* (1st ed.). New York, NY: Guilford Press.

Koo, B. W. S., Tin, A. F., Koo, E. W. K., & Lee, S. (2006). When East meets West: The implication for bereavement counseling. In C. L. W. Chan & A. Y. M. Chow (Eds.), *Death, dying and bereavement: A Hong Kong Chinese experience* (pp. 261–271). Hong Kong: University of Hong Kong Press.

McGilton, K. S., Boscart, V. M., Brown, M., & Bowers, B. (2014). Making tradeoffs between the reasons to leave and reasons to stay employed in long-term care homes: Perspectives of licensed nursing staff. *International Journal of Nursing Studies, 51*(6), 917–926. https://doi.org/10.1016/j.ijnurstu.2013.10.015.

McGilton, K. S., O'Brien-Pallas, L. L., Darlington, G., Evans, M., Wynn, F., & Pringle, D. M. (2003). Effects of a relationship-enhancing program of care on outcomes. *Journal of Nursing Scholarship, 35*(2), 151–156. https://doi.org/10.1111/j.1547-5069.2003.00151.x.

Melrose, S. (2013). Relocation stress in long term care: How staff can help. *Canadian Nursing Home, 24*(1), 16–19.

Munn, J. C., Dobbs, D., Meier, A., Williams, C. S., Biola, H., & Zimmerman, S. (2008). The end-of-life experience in long-term care: Five themes identified from focus groups with residents, family members, and staff. *The Gerontologist, 48*(4), 485–494. https://doi.org/10.1093/geront/48.4.485.

Nakrem, S., Vinsnes, A. G., & Seim, A. (2011). Residents' experiences of interpersonal factors in nursing home care: A qualitative study. *International Journal of Nursing Studies, 48*(11), 1357–1366.

National Health and Family Planning Commission of the People's Republic of China. (2014). *China's family development report 2014*. Beijing, China: China Population Publishing House [Chinese].

Nowrouzi, B., Lightfoot, N., Larivière, M., Carter, L., Rukholm, E., Schinke, R., & Belanger-Gardner, D. (2015). Occupational stress management and burnout interventions in nursing and their implications for healthy work environments: A literature review. *Workplace Health & Safety, 63*(7), 308–315. https://doi.org/10.1177/2165079915576931.

Potter, J., Stott, D. J., Roberts, M. A., Elder, A. G., O'Donnell, B., Knight, P. V., & Carman, W. F. (1997). Influenza vaccination of health care workers in long-term-care hospitals reduces the mortality of elderly patients. *Journal of Infectious Diseases, 175*(1), 1–6. https://doi.org/10.1093/infdis/175.1.1.

Pratt, C. C., Schmall, V. L., Wright, S., & Cleland, M. (1985). Burden and coping strategies of caregivers to Alzheimer's patients. *Family Relations, 34*(1), 27. https://doi.org/10.2307/583754.

Price, B. (2017). Developing patient rapport, trust and therapeutic relationships. *Nursing Standard, 31*(50), 52–63. https://doi.org/10.7748/ns.2017.e10909.

Reimer, H. D., & Keller, H. H. (2009). Mealtimes in nursing homes: Striving for person-centered care. *Journal of Nutrition for the Elderly, 28*(4), 327–347. https://doi.org/10.1080/01639360903417066.

Rickerson, E. M., Somers, C., Allen, C. M., Lewis, B., Strumpf, N., & Casarett, D. J. (2005). How well are we caring for caregivers? Prevalence of grief-related symptoms and need for bereavement support among long-term care staff. *Journal of Pain and Symptom Management, 30,* 227–233. https://doi.org/10.1016/j.jpainsymman.2005.04.005.

Rossi, A., Cetrano, G., Pertile, R., Rabbi, L., Donisi, V., Grigoletti, L., … Amaddeo, F. (2012). Burnout, compassion fatigue, and compassion satisfaction among staff in community-based mental health services. *Psychiatry Research, 200,* 933–938. https://doi.org/10.1016/j.psychres.2012.07.02.

Rusbult, C. E., & Van Lange, P. A. (2008). Why we need interdependence theory. *Social and Personality Psychology Compass, 2*(5), 2049–2070.

Ryvicker, M. (2011). Staff–resident interaction in the nursing home: An ethnographic study of socio-economic disparities and community contexts. *Journal of Aging Studies, 25*(3), 295–304. https://doi.org/10.1016/j.jaging.2010.11.001.

Skirrow, P., & Hatton, C. (2007). "Burnout" amongst direct care workers in services for adults with intellectual disabilities: A systematic review of research findings and initial normative data. *Journal of Applied Research in Intellectual Disabilities, 20*(2), 131–144. https://doi.org/10.1111/j.1468-3148.2006.00311.x.

Sreeramareddy, C. T., & Sathyanarayana, T. (2013). Decentralised versus centralised governance of health services. In The Cochrane Collaboration (Ed.), *Cochrane database of systematic reviews*. Wiley. https://doi.org/10.1002/14651858.cd010830.

Standing Committee of the National People's Congress. (2019). *The law of the people's republic of China on protection of the rights and interests of the elderly*. http://www.china.org.cn/government/laws/2007-04/17/content_1207404.htm [Chinese].

Sussmann, M., & Vecchio, R. P. (1982). A social influence interpretation of worker motivation. *Academy of Management Review, 7*(2), 177–186. https://doi.org/10.5465/amr.1982.4285547.

Thomas, W. H. (1996). The truth about nursing homes. *Life worth living: How someone you love can still enjoy life in a nursing home: The Eden alternative in action* (pp. 7–34). VanderWyk & Burnham: St Louis, MO.

Thomas, W. H., & Blanchard, J. M. (2009). Moving beyond place: Aging in community. *Generations, 33*(2), 12–17.

Wacker, R. R., & Roberto, K. A. (2013). *Community resources for elderly clients: Programs and services in an era of change*. Thousand Oaks, CA: SAGE.

Wadensten, B., & Carlsson, M. (2001). A qualitative study of nursing staff members' interpretations of signs of gerotranscendence. *Journal of Advanced Nursing, 36*(5), 635–642. https://doi.org/10.1046/j.1365-2648.2001.02019.x.

Westermann, C., Kozak, A., Harling, M., & Nienhaus, A. (2014). Burnout intervention studies for inpatient elderly care nursing staff: Systematic literature review. *International Journal of Nursing Studies, 51*, 63–71. https://doi.org/10.1016/j.ijnurstu.2012.12.001.

Wiles, J. L., Leibing, A., Guberman, N., Reeve, J., & Allen, R. E. S. (2012). The meaning of "aging in Place" to older people. *The Gerontologist, 52*(3), 357–366. https://doi.org/10.1093/geront/gnr098.

Wilson, C. B. (2012). Working together: Participation with older people. In J. Reed, C. Clarke, & A. Macfarlane (Eds.), *Nursing elderly clients* (pp. 61–76). Maidenhead, UK: Open University Press.

Wu, B., Carter, M. W., Goins, R. T., & Cheng, C. (2005). Emerging services for community-based long-term care in urban China: A systematic analysis of Shanghai's community-based agencies. *Journal of Aging & Social Policy, 17*(4), 37–60. https://doi.org/10.1300/J031v17n04_03.

Yamamoto, Y., & Holloway, S. D. (2010). Parental expectations and children's academic performance in sociocultural context. *Educational Psychology Review, 22*(3), 189–214. https://doi.org/10.1007/s10648-010-9121-z.

Yuwanich, N., Sandmark, H., & Akhavan, S. (2016). Emergency department nurses' experiences of occupational stress: A qualitative study from a public hospital in Bangkok. *Thailand. Work, 53*(4), 885–897. https://doi.org/10.3233/WOR-152181.

Zhang, Y. (2007). *The traditional Confucian viewpoint of life and death* (Unpublished doctoral thesis). Heilongjiang University, Haerbin, China.

CHAPTER 6

Mealtime Interactions in the Community

This chapter examines close and immediate interactions among elderly clients during their mealtime in the HCBS program service centers. These mealtime interactions present the micro-level interactions between elderly clients and the HCBS program. Based on literature review, we took two competing theoretical perspectives—socioemotional theory and activity theory—to examine how these mealtime interactions relate to older adults' life satisfaction. We perform secondary data analysis to test whether or not older adults' mealtime interactions in the service centers benefit their life satisfaction. Our findings suggest the value of casual, daily interactions, such as mealtime interactions, in the service centers in the HCBS program are critical for older adults with various social meanings within the micro level of the community care ecology.

6.1 Older Adults' Social Interactions

Positive social interactions can contribute to improving older adults' life satisfaction (Emlet & Moceri, 2011; Huxhold, Miche, & Schüz, 2014; McKee, Harrison, & Lee, 1999; Stacey-Konnert & Pynoos, 1992). In particular, older adults' social lives focus on interactions with immediate family members at home (e.g., Kahana, Kelley-Moore, & Kahana, 2012). Familial relationships often possess strong attachments and frequent visits, which not only provide intimate caregiving (Sneed, Whitbourne,

L. Chen and M. Ye, *Community Eldercare Ecology in China*,
https://doi.org/10.1007/978-981-15-4960-1_6

Schwartz, & Huang, 2012) but also boost self-worth and sustain older adults' well-being (Krause, 2011).

Compared with intimate interactions with family members at home or interactions based on specific purposes, such as religious services or volunteer work in the community (Okun & Schultz, 2003), social interactions in community settings tend to be casual, daily, and without a specific purpose. Such interactions have been shown to benefit older adults' well-being by expanding informal social contacts (Utz, Carr, Nesse, & Wortman, 2002), increasing social capital (Cornwell, Laumann, & Schumm, 2008), and offering emotional and instrumental support (Butler & Eckart, 2007). Studies on daily, casual interactions, such as mealtime interactions, among community-dwelling older adults who receive HCBS support, however, remain scarce. This chapter will pay special attention to older adults' social interactions on the micro or individual level, taking place in community settings (i.e., public dining rooms in the HCBS service centers).

Most previous studies investigating older adults' mealtime interactions have taken place in residential care settings, such as communal dining halls in nursing homes (Bertrand et al., 2011; Evans, Crogan, & Shultz, 2003, 2005; Nijs et al., 2006; Reimer & Keller, 2009), assisted living facilities (Chao & Dwyer, 2004; Park, 2009), and hospitals (Bryon, de Casterlé, Gastmans, Steeman, & Milisen, 2008; Simmons et al., 2013). From these studies, we know that dining together has many benefits for older adults, such as friendly talks, making new friends, and exchanging information and/or support. Curle and Keller (2010) found that the implications of mealtime interactions for older adults are so profound they can even extend to other aspects of life, such as their mental well-being and life satisfaction in general.

There may be two reasons for mealtime bringing benefits for older adults. First, dining together can increase elderly clients' food intake and nutrition (Chao & Dwyer, 2004; Evans et al., 2005; Reimer & Keller, 2009). Chao and Dwyer (2004) noted that dining with friends creates a relaxing, home-style atmosphere in assisted living facilities, which is positively associated with older adults' food intake. Evan and colleagues (2005) reported that food tasted better to elderly clients when they ate with their friends. The atmosphere of sharing food, information, emotions, and sometimes just gossip can set older adults at ease and help them better enjoy their food.

Second, dining together creates opportunities for older adults to develop new friendships and informal support networks (Gibbs-Ward & Keller, 2005; Hopper, Cleary, Baumback, & Fragomeni, 2007). This is an important means for residents in nursing homes to develop and maintain relationships (Reimer & Keller, 2009), especially when a family-style dining atmosphere is created (Altus, Engelman, & Mathews, 2002). For example, Evans et al. (2005) reported that sharing experiences when dining together helped residents to cope with the transition to a nursing home. Park (2009) found that older adults who dined together in assisted living facilities reported lower depressive symptoms than those who dined alone. Indeed, shared meals not only meet older adults' needs for nutritional intake but also create a social milieu (Gibbs-Ward & Keller, 2005).

Admittedly, not all older adults like to dine together (Bergland & Kirkevold, 2008; Curle & Keller, 2010; Evans et al., 2003). In residential care settings, some prefer to eat in their own rooms for a sense of privacy, suggesting an attempt to remain independent (Evans et al., 2003). Others are not interested in socializing during mealtimes (Bergland & Kirkevold, 2008). Curle and Keller (2010) observed conflicts among tablemates during mealtimes in residential care settings. As such, the implications of mealtime interactions for older adults' life can be equivocal. More investigation on how mealtime interactions influence older adults' lives and well-being is needed.

Also, given that most studies on mealtime interactions have taken place in residential care settings, most elderly participants in these studies were oldest-old (i.e., aged over 80 years old) or cognitively impaired (Altus et al., 2002; Aselage & Amella, 2010). Their mealtime interactions may have been largely impacted by the nursing home environment, regulations, and staff members (Keller, Laurie, McLeod, & Ridgeway, 2013). Community-dwelling older adults tend to be more physically and mentally capable than elderly residents in nursing home settings. Because HCBS service centers have almost no dining regulations—such as timing, seating, and assistance—mealtime interactions in these settings may be more organic and spontaneous than those in nursing homes. Indeed, in our interviews with service center staff members, many of them agreed that mealtime had become a pivotal social interaction occasion for elderly clients. For example, Ms. Liu pointed out:

> You know, once they do not have to cook, life becomes much easier for these older adults. It is too burdensome for them to spend a lot of time buying groceries and cooking. Eating here [in the service center], they can rest, they can talk, they can have fun in the service center. Basically, they have more time to enjoy themselves.

Another research gap relates to the sociocultural context of older adults' mealtime interactions. Existing studies on mealtime interactions among older adults primarily take place in Western countries (e.g., Evans et al., 2003; Keller et al., 2013), with few studies conducted in developing countries. Recognizing sociocultural characteristics of older adults' mealtime interactions is critical. For example, Gibbs-Ward and Keller (2005) reported that the dining setting of big, round tables in Chinese restaurants encouraged more interaction among older Canadian Chinese women than a non-Chinese dining setting did. This finding offers a cultural perspective on the atmosphere of congregate dining for Chinese older adults. Service center staff members also emphasized the importance of creating a friendly, easy-going, and family-like atmosphere for elderly clients. For example, Ms. Cao shared her view:

> Most service centers I know offer activities for elderly clients. Here, we offer health lectures weekly. We celebrate Chinese New Year with them. We have knitting, singing, dancing, and handcraft interest groups. The handcraft group is very famous. It has received many awards in the Jing'an District. The handcraft group is on Monday mornings. Tuesday afternoons, there is the knitting group. Wednesday afternoons, there is the singing group. Poker group is every Thursday. Of course, some of them just hang out here after lunch. Some of them are neighbors and some of them become good friends. They sit together every day while having lunch. Making friends through mealtime is particularly important to childless elderly clients. They need social interactions with others. That is why the HCBS program encourages a family-like approach to care for these older adults.

6.2 Community Meal Services and Health

Expanding community meal service for older adults is a growing trend in urban China. More than 81,000 older adults dined on site or used meal delivery per month across 707 service centers in Shanghai in 2017

(Shanghai Research Center on Aging, 2018), and the number of community meal service centers has increased to 722 across the city center and suburban areas in Shanghai (Shanghai Civil Affairs Bureau, 2019). In Beijing, the government has proposed to set up one community dining service for every 2000 households in the urban area and one for every 500 households in suburban areas, distributed through a food cooperative outlet (China Federation of Supply and Market Cooperation, 2019). The increasing popularity of community-based meal services suggests recognition of its importance to older adults' well-being, with profound health implications. Given the rapid increase of the aging population in Shanghai, a recent study reported that around 30% of older adults in Shanghai had at least one IADL limitations (Feng et al., 2013), including the ability to prepare meals. Meal services, thus, can help older adults in Shanghai to better age in place, maintaining their independence and autonomy (Wiles, Leibing, Guberman, Reeve, & Allen, 2012). Furthermore, dining in the community can also encourage social connections to facilitate older adults' aging in place. In the same study, Feng and his colleagues found that community-dwelling older adults in Shanghai significantly improved their ADL and IADL functions between 1998 and 2008 (Feng et al., 2013). The burgeoning of community meal services in Shanghai since 2001 may be one of the reasons, given that IADL disabilities are generally quite sensitive to contextual changes (Parker & Thorslund, 2007).

The role of HCBS programs and meal services will be pivotal in strengthening Chinese older adults' nutrition intake (Chen & Han, 2016). Is there a dining center close to home in the community? Can the dining center provide affordable and good meals? These are urgent concerns for many elderly people in Shanghai. Along with rapid economic growth in recent decades, nutritional trends, such as obesity and chronic diseases, have become significant influences on morbidity, disability, and mortality in China. The healthcare system has spent approximately 6% of China's gross domestic product (GDP) to treat these older adults for diseases such as diabetes and heart disease (Shira, 2018). This spending will likely double, adding another 4% of the entire GDP by 2025 (National Institute on Aging Demography Center, 2010). The approaching reality of multiple chronic conditions induced by unbalanced nutritional intake suggests increasing obesity and related conditions for older adults similar to the situation in Western societies. Keller (2006) found that formal meal programs, such as Meals on Wheels, can reduce in nutritional risk among community-dwelling older adults. Older adults need informal and formal supports for grocery shopping and cooking to prepare food and ensure sufficient nutritional intake (Drewnowski & Shultz, 2001). Another systematic review found that home-delivered meals have improved in quality and nutrient intake for older adults in the United States since the Older American Act of 1965 (Zhu & An, 2013). Community meal delivery service can serve also as a screening to monitor older adults' nutritional intake and malnutrition levels, as evidenced in Canada (Keller, 2007) and the United States (Dorner, 2010).

In addition, low ADL and IADL functions have been associated with malnourishment and borderline malnutrition among older adults in urban China (Han, Li, & Zheng, 2009). The Shanghai Civil Affairs Bureau (2019) regulates the quality of food in the HCBS service centers, emphasizing low sodium, low sugar, and well-balanced nutrients, as well as guaranteeing the safety of the food during the entire process of production and delivery.

In 2008, the Shanghai Civil Affairs Bureau distributed 20 million RMB (about US $3.3 million) to set up 200 additional community-based meal service centers. Shanghai had a total of 707 meal service centers in 2017 (Shanghai Research Center on Aging, 2018), all of which deliver meals and provide onsite dining.

According to building regulation policies for seniors, each service center is required to have at least 100 square meters of floor space (Shanghai Civil Affairs Bureau, 2013) and the space necessary for around 60 people to eat together (Xinhua News Agency, 2019). Some centers are more spacious; for example, the largest one has a physical space of more than 300 square meters (Shang Guan News, 2019), and another one can service 350 older adults dining together (Wen Hui Bao, 2019). Some are smaller and only a limited number of older adults can sit together for meals (Shang Guan News, 2019; Wen Hui Bao, 2019). When facing a large number of diners, staff sometimes need to divide them into different groups to accommodate the limited space available. In one case, a staff member told us that the early group starts eating at around 11:00 a.m., while the late group starts its meal at around 12:30 p.m.

The number of older adults who dine on site or have meals delivered vary greatly across service centers depending on their physical health, their housing conditions (e.g., whether their homes have elevators), and their age (e.g., according to our interviews, those over age 80 usually have meals delivered). In 2019, the Shanghai Civil Affairs Bureau set the goal of developing 1600 more community meal service centers by 2022 (Shanghai Civil Affairs Bureau, 2019). These centers will bridge the gaps between urban and rural areas in Shanghai, with the aim of extending service coverage to 50% of the total aging population (Shanghai Civil Affairs Bureau, 2019). Ultimately, the Shanghai government seeks to create a network of HCBS centers, each within a 15-minute walking distance from older adults' homes (Shanghai Civil Affairs Bureau, 2019).

6.3 Theoretical Underpinnings and Measurements

This chapter employs two competing hypotheses to examine the relationship between older adults' mealtime interactions with tablemates at the HCBS service centers and their life satisfaction. These hypotheses are based on Socioemotional Selectivity Theory (Carstensen, 1992) and Activity Theory (Cavan, Burgess, & Havighurst, 1949; Kaskie, Imhof, Cavanaugh, & Culp, 2008), respectively.

Socioemotional selectivity theory argues that as they age, older adults become more aware of their health and functional limitations and thus more selective in their interactions, strategically investing their time, energy, and resources. As a result, older adults become more likely to spend their time with significant others (i.e., their spouse, their adult children, or close friends), while they become less motivated to create new relationships in later life (Shaw, Krause, Liang, & Bennett, 2007). Some elderly clients in our study recounted their interactions with fellow clients during mealtime as "casual." For example, Mr. Wu said:

> I come and have my lunch here. If there is someone sitting beside me, I would enjoy a little talk. If I do not feel well or just do not feel like talking, I would just keep it quiet, finish my food, and go home. It's simple. I don't want to make friends [in the service centers] unless with the staff members. I'm too old for that.

As such, relationships during mealtime interactions usually do not account for intimate relationships. Based on the socioemotional selective theory, we hypothesize that:

H0 Mealtime interactions with tablemates would not be associated with older adults' life satisfaction.

In contrast, activity theory (Kaskie et al., 2008) suggests that older adults wish to create social relationships in later life. As a way to preserve self-identity in the face of age-related deficits, older adults tend to replace lost social relationships (e.g., relationships related to their previous career) with new ones (Krause, 2011). For example, Utz et al. (2002) found that widowed older adults had more informal social contacts than did their non-widowed counterparts. Fiori, Antonucci, and Cortina (2006) pointed out that networking with friends proved to be more important for predicting older adults' depressive symptoms than networking with family members (Fiori et al., 2006). More relevant to this chapter, social interactions may generate potential social support, which has been shown to have positive influences on older adults' health outcomes (Cheadle, Egger, LoGerfo, Schwartz, & Harris, 2010; Krause, 2011). In contrast to what Mr. Wu stated above, some elderly clients in our study reported that it was interesting and helpful to make a few friends in the service center. For example, Mrs. Zhao was delighted to have her social networks expanded through the meal service:

> The service center serves as a perfect social venue for our "older sisters." We all signed up for meal service and come here every day for lunch. We share the same table every noon and talk about our life, mainly our children, doctor visits, physical exercise, and so forth—you know, almost everything happening in our lives. After retirement, I did not have many people to socialize with. [In the service center], I can share my life with others with similar ages and then we become friends and we support each other. This is very important and meaningful.

Thus, in the context of this analysis, we hypothesize:

H_a More mealtime interactions would be positively associated with older adults' life satisfaction.

We used *Life Satisfaction* as the outcome variable, as measured by a Chinese version of the Health Behaviors Assessment (HBA) life satisfaction scale (Newton, Kim, & Newton, 2006). Life satisfaction is calculated by the sum of the following items: (a) In most ways my life is close to ideal; (b) The conditions of my life are excellent; and (c) I am satisfied with my life. A 5-point Likert scale ranging from 1 (strongly disagree) to 5 (strongly agree) measured responses to these three questions. This 3-item Chinese life satisfaction scale had an alpha reliability of .78.

Mealtime interaction was measured by the number of tablemates, companionship with tablemates, self-disclosure to tablemates, and instrumental support received from tablemates (Reimer & Keller, 2009). These three categories are based on Krause's (2001) multidimensional concepts of social support, which includes instrumental support (e.g., providing financial help) and emotional support (e.g., providing companionship). In this chapter, we examine how three types of social support (i.e., companionship with tablemates, self-disclosure to tablemates, and instrumental support from tablemates) received during mealtime interactions can influence older adults' life satisfaction.

Companionship here means older adults' sharing life experiences with tablemates during mealtime, such as happy memories and stories. Sharing life experiences can help to create mutual understanding and acceptance among older adults (Krause, 2011). The positive emotions evoked by these interactions can enhance older adults' physical and mental health (Burton, Mitchell, & Stride, 2011; Butler, 2006; Butler & Eckart, 2007; McKee et al., 1999).

Self-disclosure is considered an active way of seeking help, which is an important type of social support that occurs during social interactions (Krause, 2011). It involves revealing negative information about oneself to potential support providers (Henderson, Evans-Lacko, & Thornicroft, 2013; Krause, 2011). Such self-disclosure can make other people aware of older adults' needs and elicit help in time. As a result, social interactions become an efficient channel for delivering messages so potential support providers can receive the information and respond accordingly.

Instrumental support is another important resource for older adults when they have to deal with stressful events (Kahana et al., 2012; Shaw et al., 2007). Since older adults are more likely to experience physical and psychological challenges than their younger counterparts—such as bereavement and declining abilities to live independently (Cherlin, 2010; Das, 2013; DiGiacomo, Lewis, Nolan, Phillips, & Davidson, 2013; Zimmer, 2005)—offering assistance during stressful social interactions may also positively affect their life satisfaction.

The number of tablemates was measured based on respondents' reports of the number of tablemates with whom they usually dined at the service site. The answers were "I have 1 tablemate" = 1; "I have 2 tablemates" = 2; "I have 3 tablemates" = 3; "I have at least 4 tablemates" = 4.

The companionship with tablemates measure was based on respondents' responses to the question of whether they shared their life experiences or talked about their life stories during mealtimes with their tablemates. *Self-disclosure to tablemates* was measured by a single question about whether older adults talked about difficulties in their life with their tablemates. *Instrumental support from tablemates* was measured by asking whether the respondents received help from their tablemates when they were facing stressful events in their lives. Each item was measured by a 5-point Likert scale (1 = never; 2 = occasionally; 3 = frequently; 4 = very often; and 5 = always).

Our study includes control variables that previous evidence suggests may be associated with older adults' life satisfaction (Ye & Chen, 2014), including respondents' age, gender, income, education, health, and living arrangements. **Gender** was coded as a dummy variable with "female" = 1 and "male" = 0. **Age** was assessed based on years since birth. **Education** was coded as "never attended school" = 1; "primary school" = 2; "middle school" = 3; "high school" = 4; "college and/or above" = 5. **Income level** was measured by respondents' self-evaluation: "How do you evaluate your income level?" The respondents chose 1 of 5 choices: "low income" = 1; "low medium income" = 2; "medium income" = 3; "high medium income" = 4; and "high income" = 5. **Self-reported health status** was assessed by the question: "How would you describe your health?" This was rated on a 5-point Likert scale, from 1 = "very poor" to 5 = "excellent." **Living arrangements** were measured by the question: "Who are you living with?" The answers were coded into one of five categories: "living with spouse only" = 1; "living with children or grandchildren" = 2; "living with spouse and children" = 3; "living with other relatives or assistants" = 4; and "living alone" = 5.

6.4 Effects of Mealtime Interactions

Our primary analytic strategy was multilevel regression modeling to test the hypothesized relationships between mealtime interactions (i.e., number of tablemates, companionship, disclosure, and instrumental support) and older adults' life satisfaction. Multilevel regression models are useful when the data has a two-level nested structure: Level 1 is individual variables and Level 2 is group context that potentially influences Level-1 factors. In this analysis, variables at Level 1 are older adults' life satisfaction, mealtime interaction, and demographic characteristics. Service centers are the contextual variable at Level 2. By using a multilevel intercept model, we controlled the effect of service center and tested the implications of different service centers for the relationship between mealtime interaction and older adults' life satisfaction.

Level 1 model:

$$y_{i1} = \beta_{0j} + \beta_{1j}(Mealtime\ Interaction)_{ij} + \beta_{2j}(Control)_{ij} + r_{ij}.$$

Level 2 model:

$$\beta_{0j} = \gamma_{00} + \mu_{0j}$$

$$\beta_{1j} = \gamma_{10} + \mu_{1j}$$

At Level 1, older adults' life satisfaction is the sum of an intercept for service center (β_{0j}), the impacts of mealtime interactions on older adults, and older adults' demographic characteristics. The random error (r_{ij}) is associated with the ith older adult in the jth service center. At Level 2 (service center), the intercept of service centers is the sum of an overall mean (γ_{00}) and a series of random deviations from that mean (μ_{0j}). The relationship between mealtime interaction and older adults' life satisfaction is the sum of an overall slope (γ_{10}) and a series of random deviations from that slope (μ_{1j}).

Table 6.1 presents means and standard deviations for all the key variables. Among these respondents, 57% were female. Respondents were on average 74.59 years old (SD = 8.83) and had middle-school education levels (SD = 1.13). They considered themselves relatively healthy (M = 3.26; SD = 0.97) and with low medium incomes (M = 2.37; SD = 1.01). Around 33% of the respondents lived with their spouse only. The traditional multi-generational families (i.e., older adults living with their spouse, adult children, and grandchildren) accounted for 5.31% of all the respondents. Almost 32% of the respondents lived alone. On average, respondents reported relatively high satisfaction with their lives (M = 12.27; SD = 1.94). Respondents had an average of two tablemates (see Table 6.2). They enjoyed tablemates' companionship (M = 3.45; SD = 1.11), disclosed their own difficulties to their tablemates (M = 3.25; SD = 1.12), and received some instrumental support from tablemates when facing stressful events (M = 3.31; SD = 1.08). Figure 6.1 shows the prevalence of each type of mealtime interactions in percentages.

Table 6.2 shows the multilevel regression models testing the relationship between respondents' mealtime interactions and their life satisfaction.

Table 6.1 Means and standard deviations for key variables ($N = 320$)

Variables	*Mean*	*SD*
Life satisfaction	12.27	1.94
Number of tablemates	2.11	0.93
Companionship with tablemates	3.45	1.11
Disclosure to tablemates	3.25	1.12
Instrumental support from tablemates	3.31	1.08
Demography		
Female	0.57	0.50
Age	74.59	8.83
Education	3.53	1.13
Income	2.37	1.01
Health	3.26	0.97
Living arrangement		
Living with partner and children	0.05	0.22
Living with partner only	0.33	0.47
Living with child only	0.22	0.41
Living with others	0.08	0.27
Living alone	0.32	0.47

Model 1 shows that the number of tablemates had no significant association with respondents' life satisfaction. After adjusting for control variables (not shown), the number of tablemates still had no significant relationship with older adults' life satisfaction.

Models 2 through 7 show that after adjusting individual-level factors (i.e., Level 1), tablemates' companionship ($b = 0.62$; $p < .001$), disclosure to tablemates ($b = 0.47$; $p < .001$), and instrumental support from tablemates ($b = 0.45$; $p < .001$) were all positively associated with respondents' life satisfaction. The random effects on Level-2 slope in Model 3, Model 5, and Model 7 show that the influence of tablemates' companionship ($var = 0.29$; $p < .001$), disclosure to tablemates ($var = 0.44$; $p < .001$), and instrumental support from tablemates ($var = 0.40$; $p < .001$) on life satisfaction varied by service center. The statistically significant covariance between slopes and intercepts was negative. That is, for service centers whose elderly clients reported lower life satisfaction on average, the relationships of tablemates' companionship ($cov = -0.90$; $p < .001$), disclosure to tablemates ($cov = -0.97$; $p < .001$), and instrumental support from tablemates ($cov = -0.94$; $p < .001$) to older adults' life satisfaction were stronger than for service

Table 6.2 Multilevel regression for older adults' life satisfaction ($N = 320$)

Fixed-effect	*Model 1* *Coef.*	*Model 2* *Coef.*	*Model 3* *Coef.*	*Model 4* *Coef.*	*Model 5* *Coef.*	*Model 6* *Coef.*	*Model 7* *Coef.*
Number of tablemates	0.22						
Companionship with tablemates		0.68***	0.62**				
Disclosure to tablemates				0.42***	0.47***		
Instrumental support from tablemates						0.47***	0.45***
Demography							
Female			0.22		0.24		0.27
Age			−0.01		−0.01		−0.01
Education			0.02		0.03		0.03
Income			0.34***		0.35***		0.31***
Health			0.57***		0.63***		0.64***
Living arrangement *(Ref. Living alone)*							
Living with partner and children			0.76+		0.67+		0.75+
Living with partner only			−0.11		−0.21		−0.15
Living with child only			0.2		0.2		0.23
Living with others			−0.23		−1.11		−0.07
Intercept	11.70***	9.79***	6.95***	10.8***	7.21***	10.60***	7.09***
Random-Effect (var.)							
Level-2 Slope	0.47**	0.44***	0.29***	0.59***	0.44***	0.48***	0.4***
Level-2 Intercept	0.72***	1.99***	1.37***	2.54***	2.03***	2.31***	1.83***
Covariance (slope, intercept)	−0.43	−0.94***	−0.90***	−0.97***	−0.97***	−0.96***	−0.94***
Level-1 Residual	1.71***	1.61***	1.51***	1.62***	1.5***	1.65***	1.52***

(continued)

Table 6.2 (continued)

Fixed-effect	*Model 1* *Coef.*	*Model 2* *Coef.*	*Model 3* *Coef.*	*Model 4* *Coef.*	*Model 5* *Coef.*	*Model 6* *Coef.*	*Model 7* *Coef.*
Model fit							
Log Likelihood	−629.81	−611.39	−585.38	−618.13	−589.3	−620.81	−592.74
AIC	1271.63	1234.79	1200.77	1248.26	1208.59	1253.62	1215.49

+$p < .1$; *$p < .05$; **$p < .01$; ***$p < .001$

Fig. 6.1 Percentage of the degree of each type of mealtime interactions

centers whose elderly clients reported higher life satisfaction on average (Fig. 6.2).

In addition, Model 3, Model 5, and Model 7 suggest that better self-reported health status and higher income levels related to respondents' higher levels of life satisfaction. Also, compared with those who lived alone, respondents who lived in a multi-generational family enjoyed slightly higher levels of life satisfaction.

We tested two competing hypotheses about the relationship between older adults' life satisfaction and their mealtime interactions in the service centers of this HCBS program. Our findings partially support H_a, suggesting that mealtime interactions were significantly positively related with older adults' life satisfaction, while the number of tablemates was not significantly associated with life satisfaction. Daily, casual interactions with tablemates in the current HCBS program have played an important role in respondents' life. From the social-ecological point of view, everyday social interactions are the invisible and unnoticed water that constantly shapes our life (Dannefer, 2003). Despite their old age, community-dwelling respondents still actively sought social interactions.

Through a theoretical lens, respondents leaned toward H_a (hypothesis based on activity theory) but away from H_0 (hypothesis based on socioemotional selection theory). This is probably because service centers in the current HCBS program are designed to reduce transportation barriers for older adults (i.e., 15–10-minute walk on average). Most respondents who dine together are neighbors. Similar experiences and living environments can trigger conversations among older adults during mealtime (Curle & Keller, 2010). Many elderly clients and staff members participating in in-depth interviews agreed with this view. For example, Ms. Shen, a staff member, shared her view based on her daily observation:

> Elderly clients like to hang out in the service center. Because most of them grew up in this area and now still live close by, the social circle is in fact pretty small, everyone seems to know each other. Most of our clients enjoy a little chat when they have meals here. Of course, some elderly clients are more active while others keep more to themselves. More or less, these elderly clients enjoy spending time in the service center.

Mr. Qian, who had dined in the service center for approximately six years, believed that the center and the HCBS program had become his primary social network:

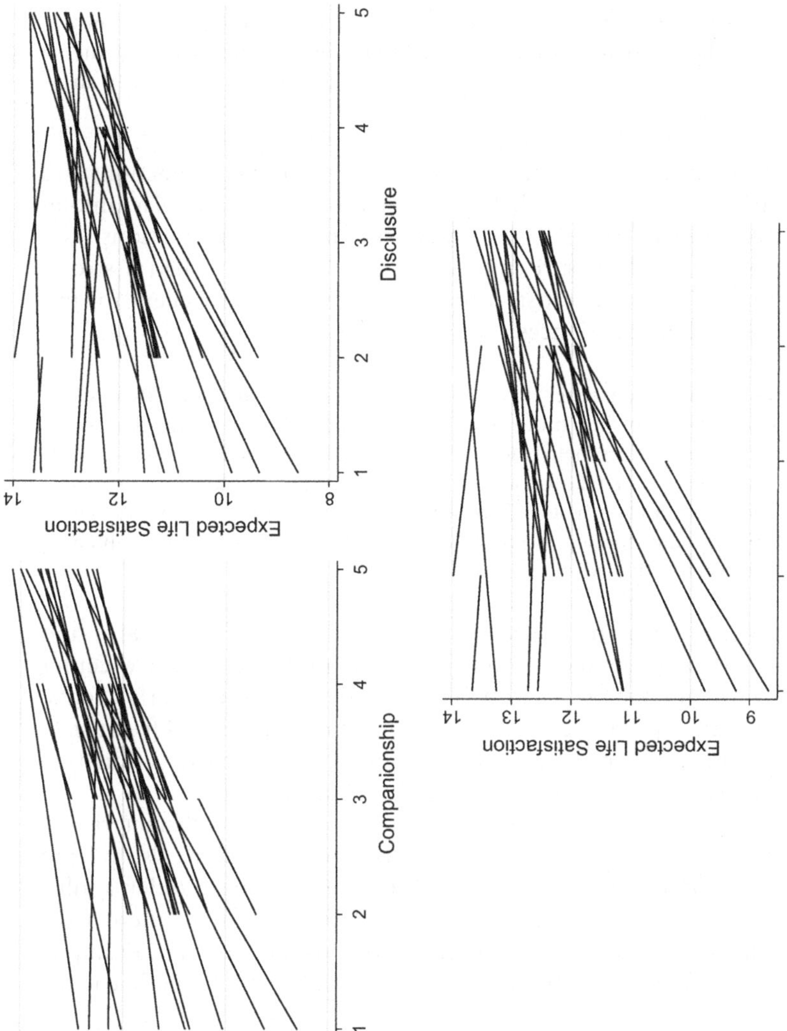

Fig. 6.2 Best-fitting slopes for the relationship between mealtime interactions and older adults' life satisfaction

> I am a widower. Staff members and friends from the HCBS program encouraged me to reconnect with society. Bereavement was so overwhelming that I was very much isolated from the outside world after my wife passed away. My neighbors told me about the meal service and brought me here. I felt welcomed here. Other older adults consoled me and helped me to let go. Since then, the service center has become an anchor of my life.

We also found that the relationship between mealtime interactions and older adults' life satisfaction varied by service centers. In particular, among older adults who had low life satisfaction, the mealtime interactions (i.e., companionship, self-disclosure, and instrumental support) had a strong positive effect on their life satisfaction. Mrs. Xue, an elderly client for five years, described her experiences:

> We check in with each other at the service center. As I am over 85 years old, I find myself relying more and more on neighbors and the service center. We older adults help each other. My children live far away, and they are all busy with taking care of their own grandchildren. My son's health condition is even worse than mine. During mealtime, we share the companionship and develop wonderful connections with each other, which is very precious at this advanced age.

Interestingly, however, relationships between mealtime interactions and life satisfaction were weaker among respondents from a service center with higher average life satisfaction. This may be related to elderly clients' other social and support networks that contributed to their life satisfaction, in which case these elderly clients might not have relied as much on mealtime interactions to seek support and companionship.

In line with earlier research, respondents' demographic characteristics, such as health status, income, and living arrangements, were also related to their life satisfaction (Ye & Chen, 2014). For Chinese older adults, dining together has a stronger sociocultural meaning than mere food intake. A majority of staff member concurred with this view. For example, Ms. Cao considered mealtime interactions the highlight of elderly clients' daily routine:

> Honestly, [elderly clients] have not much to do every day. It is good for them to dine on site, as far as I am concerned, you know, they have a place to communicate and this place is not their home. They should come out

> of their homes and communicate with people. That is, in my opinion, very important to their health.

Ms. Fang believed that service centers of the HCBS program and mealtime interactions with fellow older adults and staff members gave elderly clients a sense of belonging and being cared for:

> We boil roasted wheat tea for them every morning, and [elderly clients] come to enjoy the tea, air conditioning, and reading newspapers. It is like they come here to work every day. Ha ha! They come here every day on time. They are very happy to see us when we come here earlier than our work hours. They say "Hi!" to us cordially. They are very familiar with us. They come here, basically, for entertainment. When it is about lunchtime, we begin to prepare lunch for them. After lunch, they have various activities to participate in, like today they are playing ping-pong. These elderly clients think we are better than their children, for those who have children. You know, adult children, nowadays, only visit their elderly parents over the weekends, bringing some presents and food, acting like government officials just stopping by and going back to their own homes. They are not like us, working every day for these elderly clients, taking care of them. Do you agree?

As such, current HCBS service centers are not only for older adults who need help but widening their social networking and strengthening their sense of belonging. This finding is important because it demonstrates how micro-level interpersonal relationships (e.g., mealtime interactions) impact individuals' lives (e.g., life satisfaction), and that he influences of meso-level organizations such as HCBS service centers should also be considered (Fig. 6.1).

Individuals can influence the very environments in which they develop. Bronfenbrenner and Morris (2006) conceptualize individuals as active agents who shape their environment by instigating particular types of interactions. For example, older adults who are willing to accept other people's help are more likely to incorporate relationships with helping professionals as part of their social environments than those who go to great lengths to avoid receiving help from others. Thus, older adults who are willing to build relationships will have friendlier environments with greater resources, while those who do not will have environments that are less supportive.

Mealtime interaction network: size vs. quality. Taking a closer look at the mealtime interactions, we compared the size and quality of the interaction network and their implications for respondents' life satisfaction. In this analysis, the number of tablemates represented the network size of respondents' mealtime interactions. Companionship with tablemates, disclosure to tablemates, and instrumental support gained from mealtime interactions represented the quality of their social interactions. Prior research has shown that both network size and quality of older adults' social interactions matter (Krause, 2011).

These findings suggest that the network size of mealtime interactions (i.e., the number of tablemates) was not significantly associated with respondents' life satisfaction. This is consistent with previous evidence that the quality of social interactions is more crucial than the number of people interacting for promoting life satisfaction during old age (Gibson, 1986). One possible explanation is that larger social networks might be burdensome for older adults to maintain and may exert potentially negative impacts on their life and well-being (Stokes, 1983). This finding was also buttressed by elderly clients' responses during the in-depth interviews. For example, Mr. Yu preferred small gatherings rather than socializing with everyone in the service center. He said,

> I do not have enough energy to socialize often. Talking with several regulars during mealtime is sufficient for me. We are all old neighbors, and have known each other since childhood. These are my closest friends. There are an increasing number of older adults coming to the service center, which is a trend, but I am not interested in making any new friends. I'd like stay put with these old fellows.

Meanwhile, our analysis revealed that the quality of older adults' mealtime interactions (i.e., companionship with tablemates, disclosure to tablemates, and instrumental support from tablemates) was positively related to respondents' life satisfaction. Sharing life experience can help to create mutual understanding and acceptance among older adults (Krause, 2011). Mrs. Zhong, a 73-year-old avid volunteer, shared her story:

> Since my retirement, I began volunteering in the residential committee. I have made a lot of friends, young and old, through volunteering in the community. Some of us come to have lunch and dinner at this service center every day. It has been great for me as I started the knitting and

> crochet interest group based in this service center. We are like sisters and we help each other with anything.

The positive emotions that result from such interactions can enhance older adults' physical and mental health (Burton et al., 2011; Butler, 2006; Butler & Eckart, 2007; McKee et al., 1999). This finding echoes a recent longitudinal study that compared to the size of social networks, it is the quality of network that reduces the risk of dementia among older adults (Amieva et al., 2010). Many elderly clients we interviewed also mentioned this view. For example, Mr. Liu believed that social interactions were an antidote for dementia:

> I'm not worried at all [about dementia]. I come to [the service center] every day and chat with different people here. Mealtime offers a great opportunity for all the elderly clients to talk and socialize. Staying socially active is crucial for older adults to keep a sharp mind.

Mr. Liu shared a popular view on the relationship between social interactions and cognitive function among Chinese older adults. Social connectedness has been found to be a critical component of Chinese community-dwelling older adults' criteria for successful aging (Chen, Ye, & Kahana, 2019).

Mealtime interactions in service centers. Our analysis revealed that participation in activities at different service centers can influence both respondents' mealtime interactions and their life satisfaction. Respondents' life satisfaction varied by service center. The relationships between mealtime interactions (i.e., companionships, self-disclosures, and instrumental support) and life satisfaction also differed by service centers. In particular, for a service center where the average respondents' life satisfaction is relatively lower, the association between the social support from tablemates and their life satisfaction is significantly stronger than a service center where the average respondents' life satisfaction is higher.

Different locations of service centers may relate to this result. Older adults who live in disadvantaged neighborhoods tend to have limited access to other resources (Carpiano, 2006). Compared to their affluent counterparts, older adults from deprived neighborhoods who dined in the service centers may have encountered barriers to resources and may have proactively reached out for tablemates' support.

Another explanation may relate to staff members' work performance in different service centers. Previous research has found that staff's respect and positive attitudes enhance older adults' well-being in residential care settings (Brownie & Nancarrow, 2013; Curle & Keller, 2010). Likewise, staff in community service centers may help to foster a friendly atmosphere, encouraging older adults to spend more time there. In doing so, older adults' life satisfaction may be enhanced.

In addition, the environment of dining rooms, such as table settings (Altus et al., 2002; Nijs et al., 2006), and meal quality (Brownie & Nancarrow, 2013) may relate to older adults' mealtime interactions and further, to their life satisfaction. Service centers in the current HCBS program provide a wide range of meal plans for older adults to choose from, which is a key element of person-centered care (Reimer & Keller, 2009). Environmental characteristics beyond mealtime interactions may also influence older adults' life satisfaction (Keller et al., 2013).

6.5 Understanding Micro-Level Interactions in the Community

This chapter views interactions among older adults in service centers during their mealtime from the micro level in the social ecology (Bronfenbrenner, 1977; Lawton, Winter, Kleban, & Ruckdeschel, 1999). According to Bronfenbrenner (1977), the microsystem theory represents an environment in which individuals interact directly. It includes a collection of activities, functions, and interactions that are available to those individuals. The microsystem is the immediate social environment in which individuals live, including families and friends. Lawton et al. (1999) pointed out that a micro-level system also contains small-group environments in which individuals share similar characteristics and goals.

It is well documented that health outcomes and life satisfaction are influenced by a wide range of social system levels, including micro levels. For example, Sakellariou (2015) found that, on an individual level, seniors with disabilities have different views of their needs then those who do not have disabilities, and they required home modifications to meet these needs. Ferris, Glicksman, and Kleban (2014) found that poor housing quality had a significant impact on older adults' health, and that it could be a reason for those people to use HCBS services. The physical conditions of older adults' homes, such as dampness, heat, cold, dust, and toxins, among others, are critical for quality of life. Elderly clients who have poor housing quality tend to seek HCBS help for aging in place (Ferris et al., 2014). Moreover, service centers play an essential role in the lives of Asian older adults. Recent studies in the United States have demonstrated that Asian older adults tend to cluster together in service centers (Li, Churchill, Cheng, & Siden, 2019).

Another important finding that our study reveals is that service centers in different neighborhoods have distinct influences on the relationship between mealtime interactions and older adults' life satisfaction. For example, service centers in rich neighborhoods tend to have more resources to develop services and support elderly clients. This suggests that shared neighborhood characteristics may be a critical feature for future studies (Copping & Campbell, 2015).

In this chapter, mealtime is defined as a micro-level environment for older adults in service centers. Physically, these centers provide a dining place for elderly clients to sit, eat, and talk. According to the policy on standardized elderly activity rooms in Shanghai, each service center must have at least 100 square meters for older adults to congregate (Shanghai Civil Affairs Bureau, 2013). Some centers have a central kitchen where staff members cook for elderly clients every day. Others require outside vendors to provide meals (Shang Guan News, 2019), in which case staff members are responsible for distributing the meals to clients (People.cn, 2019). Each service center has dining tables, either square or round, so that elderly clients can sit face-to-face while dining (Shang Guan News, 2019; Wen Hui Bao, 2019). This arrangement encourages older adults to talk during their mealtime. In this micro-level environment, both older adults and staff members are actors, with older adults becoming tablemates during mealtime, sharing life experiences, and comforting each other.

Shelton (2018) pointed out that in ecological theory, each person lives in a system that consists of relationships, roles, activities, and settings. Such features of a physical environment can have a tremendous influence on people's behavior. For example, the size of the dining space in a service center is related to how many older adults can dine together. A center that can provide sufficient space for more than 100 older adults to eat together (Shang Guan News, 2019; Wen Hui Bao, 2019) creates significant opportunities for elderly clients to meet with each other and make new friends, stimulating a large number of interactions during mealtime. Some service centers are small, and only a few older adults can sit together. But the small space may also help to create an intimate atmosphere so that older adults can establish and maintain deeper, more enduring relationships during mealtime.

On the other hand, to accommodate space limitations of small service centers, elderly clients are sometimes divided into groups to have meals at different times. In that case, they do not have as many chances to interact with each other. They can, however, still interact and develop friendships within their group.

Interaction with staff is equally important. Our study also shows that the relationship between staff members and elderly clients is essential for their satisfaction with the meal service. Staff members observed that their friendship with elderly clients' is essential for the clients' quality of life. Matsui and Capezuti (2008) found that in United States, senior centers, staff members' attitudes—such as their willingness to answer questions and express concern and respect—are instrumental in ensuring fulfilling elderly clients' requests. The dining rooms of the HCBS service centers provide an environment for making friends and developing emotional and supportive connections among elderly clients and staff members during mealtime.

Our analysis also reveals that personal characteristics can affect the micro-level environment. Some elderly clients told us that because they were widowed, they came to the service center to have meals and make friends. These clients had not been significantly involved in community activities before they were widowed. Either they had been caring for their elderly spouse, or their spouse had fulfilled their need for daily interactions.

Lee and DeMaris, (2007) discovered that those who experience widowhood in their later life have often already experienced depression before losing their spouse. Ye, DeMaris, and Longmore (2018) found that people who experience widowhood report lower levels of psychological well-being and higher levels of depression than when they were married. Widowhood can cause stressful life changes, including changes in living arrangements (Kramarow, 1995), losses in social support (Ha & Ingersoll-Dayton, 2011; Lindström, 2012), and decreases in life quality (Fry, 2001; Sheykhi, 2006). In other words, because of the lack of interaction with a partner, after being widowed, surviving spouses can be faced with an emotional gap that may be difficult to fill (Anusic & Lucas, 2014). In our study, widowed elderly clients were introduced by their neighbors or friends to service centers where they could have meals and make friends. They were willing to leave their homes and go to service centers to adjust their lives (e.g., make new friends) after widowhood, and create a new life schedule. Therefore, personal characteristics, such as whether older adults are widowed or not, and whether they are willing to make friends during mealtime, are essential in determining their behaviors in the mealtime micro-level environment (Greenfield, 2012).

In addition, future studies on how micro-level environments affect community-based long-term care services are crucial. Environmental exposure and resources are situated as the backdrops of the social-ecological model (Shelton, 2018). Exposure to the environment and support can be experienced at the group level, such as having a good meal in the group setting, or at the individual level, such as chatting with other elderly clients during mealtime. The bulk of research that utilizes social-ecological theory has demonstrated that the environment of service centers can be an intervention tool, and research has also shown that interactions that aim to increase peer relationships among older adults can enhance their life satisfaction (Dickens, Richards, Greaves, & Campbell, 2011; Simoni, Franks, Lehavot, & Yard, 2011).

During mealtime, service center atmosphere also mattered to elderly clients. First, HCBS service centers should focus on environmental components that contribute to a cheerful and hospitable atmosphere to prompt communication and interactions among elderly clients. This intimate environment is vital for those who attend HCBS programs. It fosters relationships and is easy to control and tailor to the characteristics of elderly clients in the local area. Wiles et al. (2012) have shown that older adults have an intimate attachment to social spaces, where they can create their relationships, receive social support, and maintain their well-being.

Service centers should provide public services to compensate for neighborhood differences (F. Zhang, Li, Ahrentzen, & Zhang, 2019). If the service centers are in disadvantaged neighborhoods where basic public services are lacking, they may need to tailor and provide more essential services for elderly clients than centers located in affluent neighborhoods. For example, some homes do not have kitchens or bathrooms, so service centers need to provide sufficient meal services and public showers for older adults (Shang Guan News, 2019). Our study demonstrates that target subgroup features have an impact on mealtime interactions and life satisfaction. Older adults with significant financial resources, for example, were less likely to increase their life satisfaction through mealtime interactions in service centers.

Indeed, a disability-friendly environment is essential for service centers so that community-dwelling older adults' safety levels can be increased. In urban China, because the growth of the older population and the increase in life expectancy, it is expected that a higher number of older adults with functional limitations will be cared for by local communities. In November 2019, the Communist Party of China (CPC) Central Committee launched a new plan that "actively responds to the medium- and long-term planning of population aging" (CPC Central Committee, 2019). It also pointed out that public services need to prioritize economically disadvantaged and disabled older adults (CPC Central Committee, 2019).

Older adults with functional limitations often report having problems with preparing meals (Wright, Vance, Sudduth, & Epps, 2015), and nutrition is the most pressing requirement for older adults who cannot take care of themselves (Ye, Ejaz, & Rose, 2019). Studies in the United States have shown that because preparing meals can be challenging for these older adults, they tend to eat fast food with less nutrition. Another recognized problem is that older adults may eat spoiled food and poison themselves because they cannot make fresh meals on their own (Ye et al., 2019). A communal meal plan is a more efficient and cost-effective solution as opposed to hiring an assistant to cook at home for older adults (Ziegler, Redel, Rosenberg, & Carlson, 2015). Therefore, having a dining room in a HCBS service center is crucial for maintaining quality of life for older adults who need meal services. The value of using this sort of ecological thinking to aid research on community long-term care services and individual's relationships lies in its emphasis on interactions and environment, mainly focusing on context and dynamic interactions between these elements (J. Zhang, Yu, Zhang, & Zhou, 2017).

According to the policy of Shanghai government's standardization of HCBS service centers, each one is required to have disability-friendly facilities or equipment, such as mats and fall-prevention devices (Shanghai Civil Affairs Bureau, 2018). However, we found that some service centers did not have elevators for older adults because of limited budgets, which troubled some elderly clients. Henning-Smith, Shippee, and Capistrant (2018) pointed out that older adults who live alone and in poor economic conditions are more likely to be disabled. In 2016, the government of Shanghai launched a policy to promote the addition of elevators in multistory residential buildings in Shanghai (Shanghai Municipal Construction Committee, 2016). Since 2016, 700 old buildings have been renovated with newly installed elevators; however, more than 200,000 buildings still lack elevators (Shanghai Housing and Urban-Rural Construction Management Committee, 2019). It is expected that in the future, disability-friendly and fall-prevention facilities will be an important issue for policymakers and researchers to focus on.

Notably, as meal service is the most crucial and demanding service item of the HCBS program, the design of the dining room in service centers is a crucial element for micro-level interactions among elderly clients during mealtimes (Reimer & Keller, 2009). Spacious dining environments can provide opportunities for more people to sit and dine together. For example, round tables provide more opportunities for people to chat with each other while square tables allow only two to four people to talk with each other face-to-face. Peer interactions, life satisfaction, and service center physical resources are essential, and future analysis must include a consideration of these mutual influences. The impact of the general service center environment on the micro-level environment is significant. Furthermore, settings that characterize the micro-level environment are part of a larger environment (Shelton, 2018). Studies from gerontology, sociology, anthropology, and cultural studies can provide more micro-level perspectives. For example, Stokols, Lejano, and Hipp (2013) propose a social-ecological model that has spurred myriad interventions to improve health at multiple levels (Stokols et al., 2013). Intervention efforts can focus on personal interactions and the dining room environment to accommodate particular personal situations and varying levels of physical capability (Hamilton, Moore, Powe, Agarwal, & Martin, 2010).

To better facilitate micro-level interactions, staff members should be present during mealtimes to increase elderly clients' interactions. Because of limited budgets, each service center can only have two staff members during mealtime. Other staff members are busy delivering meals instead of holding engaging conversations with clients. In our study, both staff members and elderly clients at the service center commented positively on their interactions within a culture of respect for the aging and elderly (Dong & Bergren, 2016). In the future, perhaps elderly volunteers can be called upon to assist staff members in delivering meals and promoting interactions between clients and staff.

Service centers should also design and provide relevant activities before and after mealtime can also promote interactions. Mealtime is one opportunity for people to get to know each other. However, if there are more activities before or after the meal, older adults can spend more time in service centers. Providing meals is not an independent function, but one of many other functions that service centers can provide. Mealtime should be embedded within a large-scale array of social activities so that relationships and interactions among elderly customers can be continuous.

However, because personal characteristics may affect mealtime interactions, we cannot assume that each elderly client will exert the same influence or gain the same benefits from these interactions. In our study, some people who went to service centers were widowed or felt lonely. Dong and Bergren (2016) pointed out that loneliness is a crucial factor in the failure of older adults to take care of themselves. Individual characteristics and motivations may hamper or encourage the engagement of older adults, since not all people like to talk to others (Greenfield, 2012). Future interventions and policies can consider providing professional counselors to help older adults who live alone or are widowed to get out of their homes and join a service center.

Although elderly clients who live with their spouses or children who provide meals at home may not need to go to service centers to have a meal plan, it cannot be assumed that all older adults who live with their spouses or adult children do not need meal plans. Adult children may work during the day and be unable to prepare meals of their elderly parents. Traditionally, it was expected that females would stay at home and take care of older adults (Fei, 1939). Because of economic globalization and the economic reform in China since the 1980s, younger generations and more highly educated females are working outside of home (Chen, 1996). Therefore, older adults may tend to be more independent and less likely to rely on their adult children to look after them (Ye & Chen, 2014).

6.6 Chapter Summary

This chapter examined micro-level mealtime interactions among older adults in the HCBS program and how they related to their life satisfaction. It also reveals the importance of meso-level organizations on the influence of micro-level interpersonal relationships on individual elderly clients' life satisfaction. Furthermore, our findings showed that although the number of tablemates was not associated with respondents' life satisfaction, specific forms of interactions with tablemates during mealtime (i.e., companionship, self-disclosure, and instrumental support) had positive relationships with respondents' life satisfaction.

Given the variations in the influences of different service centers on older adults' life satisfaction, service programming and delivery should be tailored to environments and client characteristics of different service centers. Understanding older adults' mealtime interactions can inform the development of service delivery to prompt life satisfaction for the growing aging population in urban China.

References

Altus, D. E., Engelman, K. K., & Mathews, R. M. (2002). Using family-style meals to increase participation and communication in persons with dementia. *Journal of Gerontological Nursing, 28*(9), 47–53.

Amieva, H., Stoykova, R., Matharan, F., Helmer, C., Antonucci, T. C., & Dartigues, J.-F. (2010). What aspects of social network are protective for dementia? Not the quantity but the quality of social interactions is protective up to 15 years later. *Psychosomatic Medicine, 72*(9), 905–911. https://doi.org/10.1097/PSY.0b013e3181f5e121.

Anusic, I., & Lucas, R. E. (2014). Do social relationships buffer the effects of widowhood? A prospective study of adaptation to the loss of a spouse. *Journal of Personality, 82*(5), 367–378. https://doi.org/10.1111/jopy.12067.

Aselage, M. B., & Amella, E. J. (2010). An evolutionary analysis of mealtime difficulties in older adults with dementia. *Journal of Clinical Nursing, 19*(1–2), 33–41. https://doi.org/10.1111/j.1365-2702.2009.02969.x.

Bergland, A., & Kirkevold, M. (2008). The significance of peer relationships to thriving in nursing homes. *Journal of Clinical Nursing, 17*(10), 1295–1302. https://doi.org/10.1111/j.1365-2702.2007.02069.x.

Bertrand, R. M., Porchak, T. L., Moore, T. J., Hurd, D. T., Shier, V., Sweetland, R., & Simmons, S. F. (2011). The nursing home dining assistant program. *Journal of Gerontological Nursing, 37*(2), 34–43. https://doi.org/10.3928/00989134-20100730-04.

Bronfenbrenner, U. (1977). Toward an experimental ecology of human development. *American Psychologist, 32*(7), 513–531. https://doi.org/10.1037/0003-066X.32.7.513.

Bronfenbrenner, U., & Morris, P. A. (2006). Chapter 14. The bioecological model of human development. In W. Damon & R. M. Lerner (Eds.), *Handbook of child psychology, theoretical models of human development* (pp. 793–828). Hoboken, NJ: Wiley.

Brownie, S., & Nancarrow, S. (2013). Effects of person-centered care on residents and staff in aged-care facilities: A systematic review. *Clinical Interventions in Aging, 8,* 1–10. https://doi.org/10.2147/CIA.S38589.

Bryon, E., de Casterlé, B. D., Gastmans, C., Steeman, E., & Milisen, K. (2008). Mealtime care on a geriatric-psychiatric ward from the perspective of the caregivers: A qualitative case study design. *Issues in Mental Health Nursing, 29*(5), 471–494. https://doi.org/10.1080/01612840801981272.

Burton, E. J., Mitchell, L., & Stride, C. B. (2011). Good places for ageing in place: Development of objective built environment measures for investigating links with older people's wellbeing. *BMC Public Health, 11*(1), 839. https://doi.org/10.1186/1471-2458-11-839.

Butler, S. S. (2006). Evaluating the senior companion program. *Journal of Gerontological Social Work, 47*(1–2), 45–70. https://doi.org/10.1300/J083v47n01_05.

Butler, S. S., & Eckart, D. (2007). Civic engagement among older adults in a rural community. *Journal of Community Practice, 15*(3), 77–98. https://doi.org/10.1300/J125v15n03_05.

Carpiano, R. M. (2006). Toward a neighborhood resource-based theory of social capital for health: Can Bourdieu and sociology help? *Social Science and Medicine, 62*(1), 165–175. https://doi.org/10.1016/j.socscimed.2005.05.020.

Carstensen, L. L. (1992). Social and emotional patterns in adulthood: Support for socioemotional selectivity theory. *Psychology and Aging, 7*(3), 331–338.

Cavan, R. S., Burgess, E. W., & Havighurst, R. J. (1949). *Personal adjustment in old age*. Chicago, IL: Science Research Associates.

Chao, S., & Dwyer, J. (2004). Food and nutrition services in assisted living facilities boon or big disappointment for elder nutrition. *Generations, 28*(3), 72–77.

Cheadle, A., Egger, R., LoGerfo, J. P., Schwartz, S., & Harris, J. R. (2010). Promoting sustainable community change in support of older adult physical activity: Evaluation findings from the southeast Seattle senior physical activity network (SESPAN). *Journal of Urban Health: Bulletin of the New York Academy of Medicine, 87*(1), 67–75. https://doi.org/10.1007/s11524-009-9414-z.

Chen, L., & Han, W.-J. (2016). Shanghai: Front-runner in community-based elder care in China. *Journal of Aging & Social Policy, 28*(4), 292–307. https://doi.org/10.1080/08959420.2016.1151310.

Chen, L., Ye, M., & Kahana, E. (2019). A self-reliant umbrella: Defining successful aging among the old-old (80+) in Shanghai. *Journal of Applied Gerontology*. https://doi.org/10.1177/0733464819842500.

Chen, S. Y. (1996). *Social policy of the economic state and community care in Chinese culture: Ageing, family, urban changes, and the socialist welfare pluralism*. Aldershot, UK: Avebury.

Cherlin, A. J. (2010). *The marriage-go-round: The state of marriage and the family in America today* (Vintage edition). New York, NY: Vintage.

China Federation of Supply and Market Cooperation. (2019). *Opinions about further enhance the meal service for Chinese older adults*. Retrieved from http://www.chinacoop.gov.cn/HTML/2019/02/25/149368.html [Chinese].

Copping, L. T., & Campbell, A. (2015). The environment and life history strategies: Neighborhood and individual-level models. *Evolution and Human Behavior, 36*(3), 182–190. https://doi.org/10.1016/j.evolhumbehav.2014.10.005.

Cornwell, B., Laumann, E. O., & Schumm, L. P. (2008). The social connectedness of older adults: A national profile. *American Sociological Review, 73*(2), 185–203. https://doi.org/10.1177/000312240807300201.

CPC Central Committee. (2019, November 21). *Actively responds to the medium- and long-term planning of population aging*. http://www.gov.cn/zhengce/2019-11/21/content_5454347.htm.

Curle, L., & Keller, H. (2010). Resident interactions at mealtime: An exploratory study. *European Journal of Ageing, 7*(3), 189–200. https://doi.org/10.1007/s10433-010-0156-2.

Dannefer, D. (2003). Cumulative advantage/disadvantage and the life course: Cross-fertilizing age and social science theory. *The Journals of Gerontology Series B: Psychological Sciences and Social Sciences, 58*(6), S327–S337.

Das, A. (2013). Spousal loss and health in late life: Moving beyond emotional trauma. *Journal of Aging and Health, 25*(2), 221–242. https://doi.org/10.1177/0898264312464498.

Dickens, A. P., Richards, S. H., Greaves, C. J., & Campbell, J. L. (2011). Interventions targeting social isolation in older people: A systematic review. *BMC Public Health, 11*(1), 647. https://doi.org/10.1186/1471-2458-11-647.

DiGiacomo, M., Lewis, J., Nolan, M. T., Phillips, J., & Davidson, P. M. (2013). Health transitions in recently widowed older women: A mixed methods study. *BMC Health Services Research, 13*(1), 143. https://doi.org/10.1186/1472-6963-13-143.

Dong, X., & Bergren, S. M. (2016). The associations and correlations between self-reported health and neighborhood cohesion and disorder in a community-dwelling U.S. Chinese population. *The Gerontologist*, gnw050. https://doi.org/10.1093/geront/gnw050.

Dorner, B. (2010). Practice paper of the American dietetic association: Individualized nutrition approaches for older adults in health care communities.

Journal of the American Dietetic Association, 110(10), 1554–1563. https://doi.org/10.1016/j.jada.2010.08.023.

Drewnowski, A., & Shultz, J. M. (2001). Impact of aging on eating behaviors, food choices, nutrition, and health status. *Journal of Nutrition, Health & Aging, 5*(2), 75–79.

Emlet, C. A., & Moceri, J. T. (2011). The importance of social connectedness in building age-friendly communities. *Journal of Aging Research*, 2012. https://doi.org/10.1155/2012/173247.

Evans, B. C., Crogan, N. L., & Shultz, J. A. (2003). Quality dining in the nursing home. *Journal of Nutrition for the Elderly, 22*(3), 1–17. https://doi.org/10.1300/J052v22n03_01.

Evans, B. C., Crogan, N. L., & Shultz, J. A. (2005). The meaning of mealtimes: Connection to the social world of the nursing home. *Journal of Gerontological Nursing, 31*(2), 11–17.

Fei, H. T. (1939). *Peasant life in china*. New York: Hesperides Press.

Feng, Q., Zhen, Z., Gu, D., Wu, B., Duncan, P. W., & Purser, J. L. (2013). Trends in ADL and IADL disability in community-dwelling older adults in Shanghai, China, 1998–2008. *Journals of Gerontology. Series B, Psychological Sciences and Social Sciences, 68*(3), 476–485.

Ferris, R. E., Glicksman, A., & Kleban, M. H. (2014). Environmental predictors of unmet home-and community-based service needs of older adults. *Journal of Applied Gerontology*. https://doi.org/10.1177/0733464814525504.

Fiori, K. L., Antonucci, T. C., & Cortina, K. S. (2006). Social network typologies and mental health among older adults. *The Journals of Gerontology Series B: Psychological Sciences and Social Sciences, 61*(1), P25–P32.

Fry, P. S. (2001). Predictors of health-related quality of life perspectives, self-esteem, and life satisfactions of older adults following spousal loss: An 18-month follow-up study of widows and widowers. *The Gerontologist, 41*(6), 787–798.

Gibbs-Ward, A., & Keller, H. (2005). Mealtimes as active processes in long-term care facilities. *Canadian Journal of Dietetic Practice and Research, 66*(1), 5–11. https://doi.org/10.3148/66.1.2005.5.

Gibson, D. M. (1986). Interaction and well-being in old age: Is it quantity or quality that counts? *The International Journal of Aging and Human Development, 24*(1), 29–40.

Greenfield, E. A. (2012). Using ecological frameworks to advance a field of research, practice, and policy on aging-in-place initiatives. *The Gerontologist, 52*(1), 1–12. https://doi.org/10.1093/geront/gnr108.

Ha, J.-H., & Ingersoll-Dayton, B. (2011). Moderators in the relationship between social contact and psychological distress among widowed adults. *Aging & Mental Health, 15*(3), 354–363. https://doi.org/10.1080/13607863.2010.519325.

Hamilton, J. B., Moore, C. E., Powe, B. D., Agarwal, M., & Martin, P. (2010). Perceptions of support among older African American cancer survivors. *Oncology Nursing Forum, 37*(4), 484–493. https://doi.org/10.1188/10.ONF.484-493.

Han, Y., Li, S., & Zheng, Y. (2009). Predictors of nutritional status among community-dwelling older adults in Wuhan. *China. Public Health Nutrition, 12*(8), 1189–1196. https://doi.org/10.1017/S1368980008003686.

Henderson, C., Evans-Lacko, S., & Thornicroft, G. (2013). Mental illness stigma, help seeking, and public health programs. *American Journal of Public Health, 103*(5), 777–780.

Henning-Smith, C., Shippee, T., & Capistrant, B. (2018). Later-life disability in environmental context: Why living arrangements matter. *The Gerontologist, 58*(5), 853–862. https://doi.org/10.1093/geront/gnx019.

Hopper, T., Cleary, S., Baumback, N., & Fragomeni, A. (2007). Table fellowship: Mealtime as a context for conversation wit... : Alzheimer's care today. *Alzheimer's Care Quarterly, 8*(1), 34–42.

Huxhold, O., Miche, M., & Schüz, B. (2014). Benefits of having friends in older ages: Differential effects of informal social activities on well-being in middle-aged and older adults. *The Journals of Gerontology*. Series B, Psychological Sciences and Social Sciences, *69*(3), 366–375. https://doi.org/10.1093/geronb/gbt029.

Kahana, E., Kelley-Moore, J., & Kahana, B. (2012). Proactive aging: A longitudinal study of stress, resources, agency, and well-being in late life. *Aging & Mental Health, 16*(4), 438–451. https://doi.org/10.1080/13607863.2011.644519.

Kaskie, B., Imhof, S., Cavanaugh, J., & Culp, K. (2008). Civic engagement as a retirement role for aging Americans. *The Gerontologist, 48*(3), 368–377. https://doi.org/10.1093/geront/48.3.368.

Keller, H. H. (2006). Meal programs improve nutritional risk: A longitudinal analysis of community-living seniors. *Journal of the American Dietetic Association, 106*(7), 1042–1048. https://doi.org/10.1016/j.jada.2006.04.023.

Keller, H. H. (2007). Promoting food intake in older adults living in the community: A review. *Applied Physiology, Nutrition and Metabolism, 32*(6), 991–1000. https://doi.org/10.1139/h07-067.

Keller, H. H., Laurie, C. B., McLeod, J., & Ridgeway, N. (2013). Development and reliability of the mealtime social interaction measure for long-term care (MSILTC). *Journal of Applied Gerontology, 32*(6), 687–707. https://doi.org/10.1177/0733464811433841.

Kramarow, E. A. (1995). The elderly who live alone in the United States: Historical perspectives on household change. *Demography, 32*(3), 335–352. https://doi.org/10.2307/2061684.

Krause, N. (2001). Social support. In R. H. Binstock & L. K. George (Eds.), *Handbook of aging and the social sciences* (pp. 272–294). Academic Press.

Krause, N. (2011). Social relationships in later life. In R. H. Binstock, L. K. George, S. J. Cutler, J. Hendricks, & J. H. Schulz (Eds.), *Handbook of Aging and the Social Sciences* (pp. 181–199). Amserdam, Boston: Academic Press.

Lawton, M. P., Winter, L., Kleban, M. H., & Ruckdeschel, K. (1999). Affect and quality of life: Objective and subjective. *Journal of Aging and Health, 11*(2), 169–198. https://doi.org/10.1177/089826439901100203.

Lee, G. R., & DeMaris, A. (2007). Widowhood, gender, and depression a longitudinal analysis. *Research on Aging, 29*(1), 56–72. https://doi.org/10.1177/0164027506294098.

Li, W., Churchill, L., Cheng, J., & Siden, R. (2019, November 14). *Approaches to recruiting non-english-speaking late-life Chinese immigrants for healthy aging studies.* Gerontology Society of America 2019 Annual Scientific Meeting, Austin; TX. https://www.geron.org/images/gsa/Marketing/GSA_2019_Digital_Program_Book_v2.pdf.

Lindström, M. (2012). Marital status and generalized trust in other people: A population-based study. *The Social Science Journal, 49*(1), 20–23. https://doi.org/10.1016/j.soscij.2011.07.002.

Matsui, M., & Capezuti, E. (2008). Perceived autonomy and self-care resources among senior center users. *Geriatric Nursing, 29*(2), 141–147. https://doi.org/10.1016/j.gerinurse.2007.09.001.

McKee, K. J., Harrison, G., & Lee, K. (1999). Activity, friendships and wellbeing in residential settings for older people. *Aging & Mental Health, 3*(2), 143–152. https://doi.org/10.1080/13607869956307.

National Institute on Aging Demography Center. (2010). China's rapidly aging population. *Today's research on aging*. Washington, DC: Population Reference Bureau.

Newton, F. B., Kim, E., & Newton, D. W. (2006). A program to establish healthy lifestyle behaviors with freshmen students. *Journal of Student Affairs Research and Practice, 43*(3), 873–893. https://doi.org/10.2202/1949-6605.1681.

Nijs, K. A. N. D., de Graaf, C., Siebelink, E., Blauw, Y. H., Vanneste, V., Kok, F. J., & van Staveren, W. A. (2006). Effect of family-style meals on energy intake and risk of malnutrition in dutch nursing home residents: A randomized controlled trial. *The Journals of Gerontology*. Series A, Biological Sciences and Medical Sciences, *61*(9), 935–942.

Okun, M. A., & Schultz, A. (2003). Age and motives for volunteering: Testing hypotheses derived from socioemotional selectivity theory. *Psychology and Aging, 18*(2), 231–239. https://doi.org/10.1037/0882-7974.18.2.231.

Park, N. S. (2009). The relationship of social engagement to psychological well-being of older adults in assisted living facilities. *Journal of Applied Gerontology, 28*(4), 461–481. https://doi.org/10.1177/0733464808328606.

Parker, M. G., & Thorslund, M. (2007). Health trends in the elderly population: Getting better and getting worse. *The Gerontologist, 47*(2), 150–158.

People.cn. (2019, November 6). *Jing'an District: Developing the elderly food support service.* http://sh.people.com.cn/BIG5/n2/2019/1106/c134768-33512275.html.

Reimer, H. D., & Keller, H. H. (2009). Mealtimes in nursing homes: Striving for person-centered care. *Journal of Nutrition for the Elderly, 28*(4), 327–347. https://doi.org/10.1080/01639360903417066.

Sakellariou, D. (2015). Home modifications and ways of living well. *Medical Anthropology, 34*(5), 456–469. https://doi.org/10.1080/01459740.2015.1012614.

Shang Guan News. (2019, October 29). Memory home, happy old-age village, living committee activity circle... *Shanghai builds a happy old harbor.* https://kuaibao.qq.com/s/20191029A03NYB00?refer=spider.

Shanghai Civil Affairs Bureau. (2013, December 10). *Regulation of architecture for seniors.* http://www.shmzj.gov.cn/gb/shmzj/node8/node15/node55/node230/node262/userobject1ai7910.html.

Shanghai Civil Affairs Bureau. (2018, April 4). *The policy on construction and management of standardized elderly activity rooms in Shanghai.* http://mzj.sh.gov.cn/gb/shmzj/node8/node194/u1ai45607.html.

Shanghai Civil Affairs Bureau. (2019). *Opinions about strengthening the quality of care of meal service in Shanghai.* Retrieved from http://mzj.sh.gov.cn/gb/shmzj/node8/node194/u1ai46721.html [Chinese].

Shanghai Housing and Urban-Rural Construction Management Committee. (2019, September 9). *Survey on promoting adding elevators in old building.* http://zjw.sh.gov.cn/zjw/gzdt/20190911/71031.html.

Shanghai Municipal Construction Committee. (2016). *Adding elevators for multi-stories residential buildings in Shanghai.* http://www.shqp.gov.cn/housing/zhl/20190505/495118.html.

Shanghai Research Center on Aging. (2018). *2017 Shanghai Aging population and the development of long-term care services in 2017.* http://www.shrca.org.cn/News/detail.aspx?ID=5863&Page=0.

Shaw, B. A., Krause, N., Liang, J., & Bennett, J. (2007). Tracking changes in social relations throughout late life. *The Journals of Gerontology Series B: Psychological Sciences and Social Sciences, 62*(2), S90–S99.

Shelton, L. (2018). *The Bronfenbrenner primer: A guide to develecology.* New York: NY: Routledge.

Sheykhi, M. T. (2006). A sociological review of the reflections and dimensions of quality of life of the widows in Tehran. *Social Indicators Research, 78*(2), 251–270.

Shira, D. (2018, February 19). China's healthcare reforms underscore market growth. *China Briefing News*. https://www.china-briefing.com/news/healthcare-reforms-underscore-market-growth-china/.

Simmons, S. F., Sims, N., Durkin, D. W., Shotwell, M. S., Erwin, S., & Schnelle, J. F. (2013). The quality of feeding assistance care practices for long-term care Veterans implications for quality improvement efforts. *Journal of Applied Gerontology, 32*(6), 669–686. https://doi.org/10.1177/0733464811433487.

Simoni, J. M., Franks, J. C., Lehavot, K., & Yard, S. S. (2011). Peer interventions to promote health: Conceptual considerations. *American Journal of Orthopsychiatry, 81*(3), 351–359. https://doi.org/10.1111/j.1939-0025.2011.01103.x.

Sneed, J. R., Whitbourne, S. K., Schwartz, S. J., & Huang, S. (2012). The relationship between identity, intimacy, and midlife well-being: Findings from the Rochester adult longitudinal study. *Psychology and Aging, 27*(2), 318–323. https://doi.org/10.1037/a0026378.

Stacey-Konnert, C., & Pynoos, J. (1992). Friendship and social networks in a continuing care retirement community. *Journal of Applied Gerontology, 11*(3), 298–313. https://doi.org/10.1177/073346489201100304.

Stokes, J. P. (1983). Predicting satisfaction with social support from social network structure. *American Journal of Community Psychology, 11*(2), 141–152. https://doi.org/10.1007/BF00894363.

Stokols, D., Lejano, R. P., & Hipp, J. (2013). Enhancing the resilience of human–environment systems: A social ecological perspective. *Ecology and Society, 18*(1). JSTOR. https://www.jstor.org/stable/26269281.

Utz, R. L., Carr, D., Nesse, R., & Wortman, C. B. (2002). The effect of widowhood on older adults' social participation an evaluation of activity, disengagement, and continuity theories. *The Gerontologist, 42*(4), 522–533. https://doi.org/10.1093/geront/42.4.522.

Wen Hui Bao. (2019, November 20). *How does a highly aging community respond to the meal requirement?* http://dzb.whb.cn/2019-11-20/1/detail-661685.html.

Wiles, J. L., Leibing, A., Guberman, N., Reeve, J., & Allen, R. E. S. (2012). The meaning of "aging in place" to older people. *The Gerontologist, 52*(3), 357–366. https://doi.org/10.1093/geront/gnr098.

Wright, L., Vance, L., Sudduth, C., & Epps, J. B. (2015). The impact of a home-delivered meal program on nutritional risk, dietary intake, food security, loneliness, and social well-being. *Journal of Nutrition in Gerontology and*

Geriatrics, 34(2), 218–227. https://doi.org/10.1080/21551197.2015.1022681.

Xinhua News Agency. (2019, January 22). *Meals spots in communities will increase by 800 in 4 years to serve 160,000 people per day by 2020.* Retrieved from http://www.sh.xinhuanet.com/2019-01/22/c_137764197.htm [Chinese].

Ye, M., & Chen, Y. (2014). The influence of domestic living arrangement and neighborhood identity on mental health among urban Chinese elders. *Aging & Mental Health, 18*(1), 40–50. https://doi.org/10.1080/13607863.2013.837142.

Ye, M., DeMaris, A., & Longmore, M. A. (2018). Role of marital quality in explaining depressive symptoms after marital termination among older adults. *Marriage & Family Review, 54*(1), 34–49. https://doi.org/10.1080/01494929.2017.1284703.

Ye, M., Ejaz, F., & Rose, M. (2019). Relationships between personal needs and the environment: A study of abuse and self-neglec. In *Gerontology Society of America 2019 Annual Scientific Meeting*, Austin, TX. https://www.geron.org/images/gsa/Marketing/GSA_2019_Digital_Program_Book_v2.pdf.

Zhang, F., Li, D., Ahrentzen, S., & Zhang, J. (2019). Assessing spatial disparities of accessibility to community-based service resources for Chinese older adults based on travel behavior: A city-wide study of Nanjing, China. *Habitat International*. https://doi.org/10.1016/j.habitatint.2019.05.003.

Zhang, J., Yu, N. X., Zhang, J., & Zhou, M. (2017). Sense of community and life satisfaction in Chinese older adults: Moderating roles of personal and partner resilience. *Journal of Community Psychology, 45*(5), 577–586. https://doi.org/10.1002/jcop.21878.

Zhu, H., & An, R. (2013). Impact of home-delivered meal programs on diet and nutrition among older adults. *Nutrition and Health, 22*(2), 89–103. https://doi.org/10.1177/0260106014537146.

Ziegler, J., Redel, N., Rosenberg, L., & Carlson, B. (2015). Older Americans act nutrition programs evaluation: Meal cost analysis. *Mathematica Policy Research*. https://www.mathematica.org/our-publications-and-findings/publications/older-americans-act-nutrition-programs-evaluation-meal-cost-analysis.

Zimmer, Z. (2005). Health and living arrangement transitions among China's oldest-old. *Research on Aging, 27*(5), 526–555. https://doi.org/10.1177/0164027505277848.

CHAPTER 7

Conclusion

We have examined how HCBS influences community care ecology for older adults in Shanghai. Through the HCBS program, older adults interact with the macro, meso, and micro levels of the community ecology, respectively. Informed by community gerontology and the social-ecological framework, this chapter reviews the previous chapters and illustrates how the HCBS program encompasses and unites these three levels, producing a community care ecology for older adults. The chapter concludes with suggestions for future directions in eldercare policy and practice development in Shanghai.

7.1 Aging in Community

In previous chapters, we analyzed multidimensional and dynamic interactions between elderly clients and program staff in the Shanghai HCBS program using the community gerontology model Greenfield, Black, Buffel, and Yeh (2019) developed. According to this model, there are four pathways nested around the meso level in the community: (a) meso-level influences on the micro level, (b) micro-level influences on the meso level, (c) macro-level influences on the meso level, and (d) meso-level influences on the macro level (Greenfield et al., 2019). Following Greenfield et al., we conceptualize the HCBS program as a meso-level construct within the Shanghai community care ecology.

L. Chen and M. Ye, *Community Eldercare Ecology in China*,
https://doi.org/10.1007/978-981-15-4960-1_7

Meso-level influences on the micro level. We analyzed mealtime interactions in the HCBS service centers and found that onsite dining services are a vital source of elderly clients' interpersonal interactions. For example, in Chapters 5 and 6, we discussed how environmental preferences influenced older adults' mealtime interactions. It is important to recognize that HCBS service centers foster interpersonal interactions for their elderly clients, who come to the centers every day for lunch and mealtime activities. In particular, regular mealtime interactions provide an environment in which older adults can share their positive and negative life experiences. As Shaw, Krause, Liang, and Bennett (2007) suggested, older adults may share their feelings and experiences only within a very close circle, because their social networks shrink as they age. But living in the same neighborhood creates opportunities for older adults to interact and share their feelings (Shaw et al., 2007). For example, Curle and Keller (2010) found that having shared community and life experiences can facilitate older adults' conversations during mealtime. Congregated dining environments can also hold strong sociocultural meanings for older adults who dine together in long-term care settings (Gibbs-Ward & Keller, 2005). The HCBS program should use mealtimes to enhance familiarity among its elderly clients, which may improve their life satisfaction (Chen, Ye, & Kahana, 2019a). Indeed, arranging opportunities for mealtime socializing has been found to improve older adults' reception of HCBS services in both Chinese and Western contexts (Chao, Houser, Tennstedt, Jacques, & Dwyer, 2007; Chen et al., 2019a; Reimer & Keller, 2009).

HCBS staff also offer an important opportunity for elderly clients to socialize at the service centers. According to Donabedian (1988) as well as Kajonius and Kazemi (2015), frequent interpersonal interactions facilitate service delivery and improve quality of care. At the Shanghai HCBS facilities, staff members had daily interactions with older adults, which enabled them to provide responsive service. Staff mentioned the importance of treating elderly clients with respect during service hours and reported being satisfied with their jobs when older adults responded positively to their care. The Shanghai staff also recognized the importance of protecting elderly clients' dignity, and insisted on delivering tailored services to help those who were in need. In Chinese culture, filial piety—"Respect the old and love the young" (*zun lao ai you* in Chinese)—is not only observed in families but also touches every aspect of social life (Ikels, 2004). Despite their limited workforce, HCBS program staff endeavored to interact with as many elderly clients as they could during mealtime. In doing so, staff members followed the HCBS program's mission to provide a family atmosphere while delivering care.

Also, it is essential for HCBS service centers to offer clients a variety of social activities. In Shanghai, HCBS staff devised a variety of activities to promote social interactions among elderly clients, expanding their clients' social networks and strengthening their sense of community. Kojola and Moen (2016) found that a familiar community can engender a sense of security and comfort for older adults during and after retirement, supporting their psychological and role transitions (Kim & Moen, 2002). Chen, Ye, and Kahana (2019b) likewise reported that even for the old-old population (80 years old and above) in Shanghai, the HCBS program can be a valuable venue in which elderly clients develop a sense of community (Chen et al. 2019b).

Of course, meso-level characteristics exist not only within the service centers or the HCBS program itself; neighborhood and environmental considerations should also be analyzed (Greenfield et al., 2019). Accessibility to various community-based services, including HCBS; housing conditions; neighborhood construction; income; and other meso-level characteristics should be integrated throughout the entire HCBS system. For instance, as staff members suggested, the HCBS program could hold regular sessions promoting health, education, and volunteering opportunities across sub-districts to keep older adults engaged with their community.

Micro-level influences on the meso level. Given the increased time older adults spend in their neighborhood after retiring, the neighborhood becomes the most immediate and convenient location for them to access eldercare services and support (Buffel et al., 2012). Through interpersonal interactions on the micro level of the community ecology, older adults seek emotional and instrumental support from the HCBS program establish community among an extended interpersonal network (Buffel et al., 2012; Dury et al., 2014).

Buffel and his colleagues (2012) drew on the Convoy Model (Fiori, Antonucci, & Akiyama, 2008) to demonstrate the dynamic social ties among older adults in the same community, ranging from family members and close friends to neighbors and acquaintances (Buffel et al., 2012). In our study, HCBS program staff formed a pivotal convoy for community-dwelling older adults in Shanghai. Staff interact with older adults on a daily basis, providing instrumental support, emotional comfort, as well as program information. Because staff recognized the importance of providing tailored services to meet older adults' requirements and needs (Chen, Ye, & Kahana, 2018), they listened to their elderly clients' feedback and responded to their questions and requests. Staff also performed services that went beyond the scope of their job responsibilities, including monitoring blood pressure, coordinating day trips, and helping clients shop online. These extra accommodations improved older adults' perception of HCBS, and could gradually be incorporated into the program, thereby improving quality of care and ensuring that employees receive proper compensation.

While it is widely accepted that strict regulatory caregiving standards should not supersede relationships with elderly clients (Chung, 2012) or be an impediment to meeting their actual needs (Putnam, Tang, Brooks-Danso, Pickard, & Morrow-Howell, 2007), HCBS staff lack institutional guidelines that outline professional conduct. Currently, staff members follow cultural traditions—filial piety, in particular—to induce a family-like atmosphere when administering services. Because some elderly clients asked for assistance beyond what the HCBS program provided, staff had to assess if they were allowed to help these elderly clients, adding further duress to their workload. Filial piety mandates that younger generations follow the instructions of their older adult relatives, but it also burdens staff members with unreasonable demands, especially when their elderly clients do not have children of their own. As older adults' actions can influence and predict staff members' behaviors and emotions (Kenny, Kashy, Cook, & Simpson, 2006), staff frequently worried that strained relationships could impede their ability to provide care, which further influenced how they served their HCBS clients.

Chung (2012) found that caregiving relationships in nursing homes could be too nuanced to be standardized, which was often overlooked during service delivery. Our findings suggest that caregiving relationships in the HCBS contexts can be even more so. HCBS programs should establish detailed guidelines that protect staff while fostering caregiving relationships with clients and clarifying how services should be delivered—especially when elderly clients require personal care or assistance. When HCBS staff members and elderly clients have agency over their own decision-making, they can actively determine how they will participate in or modify program guidelines, which suggests that micro-level interactions can have an impact on the meso level.

Macro-level influences on meso level. In China, rapid urbanization has caused unprecedented changes in the expression of family values, influencing public perceptions of eldercare (Holroyd, 2003; Whyte, 2004). In particular, the Economic Reform has increased the number of migrant workers in the country and lessened the popularity of traditional family caregiving models. Also, the dissolution of filial piety, has given younger generations abundant material and financial advantages (Ikels, 2004; Yu, 2013) and the ability to decide whether or not they will be their elderly parents' caregivers (Chen, 2016).

HCBS programs emerged in Shanghai as a response to the evolving state of Chinese eldercare (Chen, 2016). In response to the rapidly increasing number of Chinese aging population, HCBS has become a viable alternative to traditional family caregiving (Z. Feng, Liu, Guan, & Mor, 2012). In Chapter 3, we reviewed the history of HCBS development in Shanghai since 2007 till now. During this development, the proposal of the "90-7-3" eldercare framework highlighted HCBS (Shanghai Government, 2007). Guided by the "90-7-3" framework, HCBS programs in Shanghai aim to centralize geriatric and long-term care resources to support the majority of its older residents (i.e., 97%) who do not require assisted living or institutional care.

Because older adults may find it difficult to navigate services and policies, such as health and social services, financial reimbursement and entitlement, housing renovation, and so forth (Greenfield et al., 2019), the HCBS service centers in our study provided a location in which older adults receive a variety of services and support. In Chapter 5, we illuminated how staff members disseminated information about the HCBS program—for example, telling older adults when and where to receive service coupons or their entitled service reimbursements—in addition to providing instrumental support, such as meal services. By providing this information while serving clients, HCBS staff expedited the process by which their elderly clients received benefits and support.

Since its introduction, HCBS has become increasingly popular in Shanghai, which has enabled older adults in the city to recognize its value. In a qualitative study, Chen et al. (2019b) found that HCBS helped older adults living in Shanghai retain their self-reliance and social connectedness, thereby becoming a key component in their definition of successful aging. Similarly, Q. Feng et al. (2013) conducted a population-based survey that suggested older adults in Shanghai had improved significantly in both ADL and IADL functions over the decade of 1998–2008. Because older adults' IADL functions are generally more sensitive to contextual changes than their ADL functions (Parker & Thorslund, 2007), the development of HCBS in Shanghai might have contributed to improvements in their quality of life and well-being.

At the same time, the burgeoning economy in Shanghai has strained HCBS development in the Jing'an District. In Chapter 4, we discussed how soaring real estate prices in Shanghai have challenged the HCBS program's expansion. The cost of acquiring new service centers and renovating existing centers rose before the government's budget could accommodate these shifts.

On the macro level, widespread social change continues to influence the development of HCBS programs, but innovations in community-based eldercare in urban China remain undertheorized and underdeveloped. It is crucial for future researchers to investigate the implications of social, cultural, and historical change when researching aging populations in China.

Meso-level influences on the macro level. The Jing'an District HCBS program initially began with food preparation for older adults living in the Jing'an Temple sub-district. While interviewing elderly clients receiving HCBS services in the district, we learned that time-consuming food preparation—including grocery shopping, cooking, and cleaning—had posed unexpected difficulties as their age advanced.

In 2007, the Shanghai Civil Affairs Bureau (2008) conducted a small-scale survey, which estimated that 23.7% of community-dwelling older adults in Shanghai wished to receive meal service, either by delivered meals or by dining in congregated settings. In 2008, the Shanghai Civil Affairs Bureau distributed 20 million yuan (roughly US$3 million) to set up 200 community-based dining service centers, including two types: (a) comprehensive, including food preparation, delivery, and dining in, for at least 150 people per meal per day; and (b) dining in only, for no fewer than 50 people per meal per day (Shanghai Civil Affairs Bureau, 2008). In Chapter 5, we discussed how staff members routinely pointed out that meal service has steadily increased over the years and has remained the service most in demand among the 10 service types the HCBS program offers. After introducing the meal service in 2007, the Shanghai HCBS program began combining previously distinct services and developing new ones. The Shanghai government's response to meal service turned out to be the catalyst for the burgeoning HCBS in the city. In Chapter 4, we argued that site selection and resource allocation for service centers were related to neighborhood conditions, which were in turn governed by older adults' ability to walk to their local service center. We likewise determined that older adults' requests and needs could affect HCBS development.

Indeed, older adults should have input not only in health- and service-related matters but also in community policymaking (Buffel et al., 2012). For example, for HCBS centers to best satisfy their elderly clients' needs, they should tailor their services to the most pressing needs of their local constituents (Buffel, Verté, De Donder, & De Witte, 2008). These age-friendly, meso-level efforts encourage the evolution of HCBS programming and the development of further, relevant policies on the macro level (Greenfield et al., 2019; Greenfield, Oberlink, Scharlach, Neal, & Stafford, 2015). HCBS staff who believe their efforts will improve future HCBS services and programming demonstrate that early investment in community-based solutions leads to more effective eldercare.

Time and generation. Time is an essential construct when theorizing community and aging (Greenfield et al., 2019). Elder, Shanahan, and Jennings (2015), outlining a life-course perspective, contend that individuals' cumulative life experiences influence their later life, using the Chinese baby-boomer generation as a particular example (Elder et al., 2015).

Between 1953 and 1964, the Chinese population swelled by 112 million (Riskin, 2000). The children born during the population influx of the 1950s and 1960s are now referred to as the Chinese baby-boomer generation (Greenhalgh, 2005). As this generation begins to age, China faces a bloating population of elders and middle-aged adults in the approaching decade (Du, 2013). Because some Chinese baby boomers have already reached retirement age, it is critical to recognize the role and meanings of community and HCBS for this generation—especially because most of the Chinese baby boomers are the parents of only children required by the one-child policy who will face a significant shortage of family caregivers in the near future. Also, the identity of Chinese baby boomers arose from a series of significant historical events and policy changes across their life courses, which played a critical role in shaping their perceptions of filial piety, long-term care, and family caregiving (Chen, 2016; Liang, 2011). In Chapter 5, we reported on interviews with HCBS staff who agreed that they would likely use HCBS services when they reached an advanced age to avoid burdening their only child. Even the old-old in Shanghai have recognized the changing social and cultural contexts of caregiving and have begun to accept HCBS (Chen et al., 2019b).

HCBS in Shanghai should acknowledge that older adults do not constitute a homogeneous group (Buffel et al., 2012; Greenfield et al., 2019). Age gaps among care recipients should inform our constructs of aging and community. The young-old group (i.e., 60–75 years old) may prefer more social activities, volunteering, civic engagement, and expanding their social networks, while the old-old group may need more assistance and support to remain autonomous. Implementation of HCBS programs and policy should not only focus on current older adults' needs but also anticipate the young-old generation's potential needs, identifying shifting generational characteristics, because older adults' needs and adaptations are best understood in the context of their unique life courses (Elder et al., 2015; Scheidt, Humpherys, & Yorgason, 1999).

7.2 Implications for Policy and Practice

Our findings suggest a range of policy implications for health care in the community ecology, using HCBS in urban Shanghai as an example. Chinese healthcare policymakers should first acknowledge an urgent fact: China's aging population is not only growing rapidly but its definitions of community and communal responsibility are changing. China has experienced major societal change along with the rapid growth of its aging population, which has significantly increased their life expectancy (Zeng, Feng, Hesketh, Chistensen, & Vaupel, 2017). As the "oldest city" in China, Shanghai houses the highest proportion of the old-old population and boasts the highest life expectancy rate nationwide (83.37 years old in 2017; Shanghai Research Center on Aging, 2018). The old-old population in Shanghai has increased by 60% in a decade, from about 502,000 in 2007 to over 805,000 in 2017, accounting for 16.7% of the total 60-years-and-older population in China (Shanghai Research Center on Aging, 2008, 2018). Providing eldercare and infrastructure for urban China's larger aging population thus assumes major significance. Taking the aforementioned differences between the young-old and old-old into account, HCBS policies need to anticipate generational changes that will affect how the programs implement their services. We believe an array of generationally specific service items and delivery methods should be devised to meet the needs and requests of unique age and regional groups among Chinese older adults.

To better cater to the varying needs of this growing aging population, Chinese healthcare policies should continue to underscore the role of community in eldercare. Given the declining prevalence of family caregiving and the implications of the one-child policy, it is likely that community-dwelling older adults in urban China will seek personal assistance and social support from HCBS programs. Adapting to this shift in caregiving standards, policymakers should embrace timely and necessary program expansions tailored to discrete aging populations. As Buffel et al. (2012) note, community is both a geographical location and a social environment for older adults; it is natural for aging adults to seek security and safety in their local community. Chen et al. (2019b) likewise found that living in a familiar environment can help community-dwelling older adults live as long as possible and maintain their self-reliance. As such, HCBS policies should acknowledge that community is in fact "a multilayered and spatially contextualized process" (Buffel et al., 2012, p. 21). Relevant policy design and implementation must foreground interrelated variables, such as instrumental and social support, to avoid isolating elderly clients seeking HCBS services.

From a macro-level perspective, Chinese HCBS policymakers should continually revisit the age-friendly initiative the World Health Organization (WHO) published in 2007. Greenfield et al. (2015, p. 192) recommend that age-friendly cities make "deliberate and distinct efforts across stakeholders from multiple sectors within a defined and typically local geographic area to make social and/or physical environments more conducive to older adults' health, wellbeing, and ability to age in place and in the community." The Shanghai government published its own age-friendly city guidelines in 2013 (Shanghai Civil Affairs Bureau, 2013). The Shanghai guidelines, following most of the WHO's directions, range from topics such as public space and transportation to individual older adults' housing and community. In particular, the Shanghai guidelines highlight the significance of the HCBS program when assisting older adults' aging in community (Shanghai Civil Affairs Bureau, 2013). Community-dwelling older adults in urban China should be further empowered to address their unmet needs within their community and family (Rémillard-Boilard, Buffel, & Phillipson, 2017), discuss local environmental conditions that could affect service delivery or the development of new services (Giunta & Thomas, 2015), and advocate for themselves, especially during drastic social change (Buffel, Handler, & Phillipson, 2018).

HCBS, the meso construct in the community care ecology, should be the intersection of the macro and micro levels to develop better support and welfare for older adults (Greenfield et al., 2019). Because the majority of older adults will age at home and in the community, awareness of community ecology can orient various policy efforts meant to improve aging-related HCBS programs.

In addition to policy implications, our findings offer practical suggestions for ways to better design and implement HCBS programming. On the macro level of the community ecology, HCBS programming should empower older adults to determine their own needs and concerns, especially with respect to service delivery, quality of care, and new service development. Proactively educating older adults about health care and HCBS programs can encourage them to engage with their local community. A healthy community ecology, which spans intertwined physical, social, organizational, and cultural environments, is essential for older adults' well-being and life satisfaction (Lawton, 1982). HCBS programs in Shanghai should acknowledge such multifaceted characteristics of community ecology for older adults and begin to extend their service items from personal assistance to broader social components, such as volunteering and civic engagement. To serve the widest possible array of older adults, HCBS programs should consider resource allocation and location selection when determining how to bundle their eldercare service items.

On the meso level of community ecology, integrating different service items should be another primary concern for the HCBS programs in Shanghai. Service centers should seek opportunities for intra-HCBS collaboration that would not only promote their healthcare services but increase older adults' social engagement and expand their social networks. Local development and changes offer valuable opportunities for practitioners and social workers to tailor their services to the needs of older adults living in the neighborhood. Also, preemptively engaging older adults in the conversation about HCBS development is a useful and effective way to advance quality of care (Black & Lipscomb, 2017). Older adults' can articulate their own needs and requests more accurately than HCBS staff members or social workers; listening to their feedback can help HCBS provide corresponding services and locate necessary resources. HCBS programs in Shanghai should also emphasize the value of social interaction. Especially when older adults are in good physical health, receiving social support from their community has been found to significantly lower their stress levels and improve successful aging in Western contexts (Bowling & Stafford, 2007; Moore et al., 2015).

Community-dwelling older adults in urban China also report that social interaction and social networks are key components of their successful aging (e.g., Chen et al., 2019b; Li et al., 2014). Indeed, older adults can educate community developers on an informal basis (Buffel et al., 2012).

On the micro level, in addition to promoting interactions among older adults, HCBS in Shanghai should address the needs of its staff members. It is crucial for HCBS programs to provide a list of external resources so staff members can increase their efficiency, decrease their uncompensated labor, and connect clients with other community-based programs. In particular, HCBS staff members should receive enough labor hours and corresponding adequate compensation to familiarize themselves with the elderly clients they care for (Westermann, Kozak, Harling, & Nienhaus, 2014). More professional training could be another cost-effective way to improve their ability to meet elderly clients' needs. Current HCBS regulations should provide refined, specific protocols for how each step or service should be administered.

In sum, HCBS practice and programming in Shanghai should focus on community-level changes and opportunities (Greenfield et al., 2019). Service centers should view the community as a dynamic and holistic organism that includes older adults. This organism possesses its own ecology that hosts the intertwined micro, meso, and macro levels.

7.3 Book Summary

Community Care Ecology in China conceptualizes community as an ecology that hosts HCBS programs and a variety of environments for older adults in Shanghai. We reviewed the historical development of HCBS in Shanghai, the allocation of service centers, staff members' experiences, and elderly clients' interpersonal interactions across the macro, meso, and micro levels of community ecology. HCBS has become the nexus for older adults seeking to fulfill their personal, instrumental, and social needs. Following the social-ecological framework, we illustrated current HCBS progress in Shanghai and suggests directions for its future development. Aging in community goes beyond aging in place, integrating the multidimensional, holistic aspects of their environment for older adults. Given the rapid aging trend in China, the meanings of HCBS should be further extended and enriched as a key element in the community ecology for older adults.

References

Black, K., & Lipscomb, V. B. (2017). The promise of documentary theatre to counter ageism in age-friendly communities. *Journal of Aging Studies, 42*, 32–37. https://doi.org/10.1016/j.jaging.2017.06.001.

Bowling, A., & Stafford, M. (2007). How do objective and subjective assessments of neighbourhood influence social and physical functioning in older age? Findings from a British survey of ageing. *Social Science and Medicine, 64*, 2533–2549. https://doi.org/10.1016/j.socscimed.2007.03.009.

Buffel, T., Handler, S., & Phillipson, C. (2018). Age-friendly cities and communities: A manifesto for change. In T. Buffel, S. Handler, & C. Phillipson (Eds.), *Age-friendly cities and communities: A global perspective* (pp. 273–288). Chicago, IL: The University of Chicago Press.

Buffel, T., Verté, D., De Donder, L., & De Witte, N. (2008). Exploring the links between conceptions of the neighbourhood and gerontological research perspectives. In A. Balci, C. Aktan, & O. Dalbay (Eds.), *Studies on social sciences*. Izmir: Social Sciences Research Society.

Buffel, T., Verté, D., De Donder, L., De Witte, N., Dury, S., Vanwing, T., & Bolsenbroek, A. (2012). Theorising the relationship between older people and their immediate social living environment. *International Journal of Lifelong Education, 31*(1), 13–32. https://doi.org/10.1080/02601370.2012.636577.

Chao, S. Y., Houser, R. F., Tennstedt, S., Jacques, P., & Dwyer, J. T. (2007). Food and nutrition care indicators: Experts' views on quality indicators for food and nutrition services in assisted-living facilities for older adults. *Journal of the American Dietetic Association, 107*(9), 1590–1598. https://doi.org/10.1016/j.jada.2007.06.006.

Chen, L. (2016). *Evolving eldercare in contemporary China: Two generations, one decision*. New York, NY: Springer.

Chen, L., Ye, M., & Kahana, E. (2018). "Their today is our future": Direct carers' work experience and formal caring relationships in a community-based eldercare program in Shanghai. *Journal of Applied Gerontology, 37*, 516–537. https://doi.org/10.1177/0733464816653360.

Chen, L., Ye, M., & Kahana, E. (2019a). Process and structure: Service satisfaction and recommendation in a community-based elderly meal service in Shanghai. *Journal of Aging & Social Policy*. https://doi.org/10.1080/08959420.2019.1704132.

Chen, L., Ye, M., & Kahana, E. (2019b). A self-reliant umbrella: Defining successful aging among the old-old (80+) in Shanghai. *Journal of Applied Gerontology*. https://doi.org/10.1177/0733464819842500.

Chung, G. (2012). Nursing assistant views on nursing home regulatory inspection: Knowledge and attitudes regarding the state nursing home survey. *Journal of Applied Gerontology, 31*(3), 336–353.

Curle, L., & Keller, H. (2010). Resident interactions at mealtime: An exploratory study. *European Journal of Ageing, 7*(3), 189–200. https://doi.org/10.1007/s10433-010-0156-2.

Donabedian, A. (1988). The quality of care: How can it be assessed? *JAMA, 121,* 1145–1150.

Du, P. (2013). Intergenerational solidarity and old-age support for the social inclusion of elders in Mainland China: The changing roles of family and government. *Ageing & Society, 33,* 44–63. https://doi.org/10.1017/s0144686x12000773.

Dury, S., Willems, J., De Witte, N., De Donder, L., Buffel, T., & Verté, D. (2014). Municipality and neighborhood influences on volunteering in later life. *Journal of Applied Gerontology, 35*(6), 601–626. https://doi.org/10.1177/0733464814533818.

Elder, G. H., Shanahan, M. J., & Jennings, J. A. (2015). Human development in time and place. In M. H. Bornstein and T. Leventhal (Eds.), *Handbook of child psychology and developmental science* (Vol. 4, 6th ed., pp. 665–715). New York, NY: Psychology Press.

Feng, Q., Zhen, Z., Gu, D., Wu, B., Duncan, P. W., & Purser, J. L. (2013). Trends in ADL and IADL disability in community-dwelling older adults in Shanghai, China, 1998–2008. *Journals of Gerontology, Series B: Psychological Sciences and Social Sciences, 68*(3), 476–485. https://doi.org/10.1093/geronb/gbt012.

Feng, Z., Liu, C., Guan, X., & Mor, V. (2012). China's rapidly aging population creates policy challenges in shaping a viable long-term care system. *Health Affairs, 31,* 2764–2773. https://doi.org/10.1377/hlthaff.2012.0535.

Fiori, K., Antonucci, T. C., & Akiyama, H. (2008). Profiles of social relations among older adults: A cross-cultural approach. *Ageing & Society, 28,* 203–231.

Gibbs-Ward, A., & Keller, H. (2005). Mealtimes as active processes in long-term care facilities. *Canadian Journal of Dietetic Practice and Research, 66*(1), 5–11. https://doi.org/10.3148/66.1.2005.5.

Giunta, N., & Thomas, M. L. (2015). Integrating assessment and evaluation into partnership initiatives: Lessons from the community partnerships for older adults. *Journal of Applied Gerontology, 34*(5), 609–632. https://doi.org/10.1177/0733464813487587.

Greenfield, E. A., Black, K., Buffel, T., & Yeh, J. (2019). Community gerontology: A framework for research, policy, and practice on communities and aging. *The Gerontologist, 59,* 803–810. https://doi.org/10.1093/geront/gny089.

Greenfield, E. A., Oberlink, M., Scharlach, A. E., Neal, M. B., & Stafford, P. B. (2015). Age-friendly community initiatives: Conceptual issues and key questions. *The Gerontologist, 55,* 191–198. https://doi.org/10.1093/geront/gnv005.

Greenhalgh, S. (2005). Missile science, population science: The origins of China's one-child policy. *The China Quarterly, 182,* 253–276.

Holroyd, E. E. (2003). Chinese family obligations toward chronically ill elderly members: Comparing caregivers in Beijing and Hong Kong. *Qualitative Health Research, 13*(3), 302–318. https://doi.org/10.1177/1049732302250127.

Ikels, C. (Ed.). (2004). *Filial piety: Practice and discourse in contemporary East Asia.* Standford, CA: Stanford University Press.

Kajonius, P., & Kazemi, A. (2015). Structure and process quality as predictors of satisfaction with elderly care. *Health and Social Care in the Community, 24*(6), 699–707. https://doi.org/10.1111/hsc.12230.

Kenny, D. A., Kashy, D. A., Cook, W. L., & Simpson, J. A. (2006). *Dyadic data analysis* (1st ed.). New York, NY: Guilford Press.

Kim, J. E., & Moen, P. (2002). Retirement transitions, gender, and psychological well-being: A life-course, ecological model. *Journal of Gerontology: Psychological Sciences, 57B*(3), P212–P222.

Kojola, E., & Moen, P. (2016). No more lock-step retirement: Boomers' shifting meanings of work and retirement. *Journal of Aging Studies, 36,* 59–70.

Lawton, M. P. (1982). Competence environmental press, and the adaptation of older people. In P. Lawton, P. Windley, & T. Byerts (Eds.), *Aging and the environment* (pp. 33–59). New York, NY: Springer.

Li, C. I., Lin, C. H., Lin, W. Y., Liu, C. S., Chang, C. K., Meng, N. H., … Lin, C. C. (2014). Successful aging defined by health-related quality of life and its determinants in community-dwelling elders. *BMC Public Health, 14*(1), 1013–1021.

Liang, J. Y. (2011). Components of a meaningful retirement life: A phenomenological study of the 1950s birth cohort in urban China. *Journal of Cross-Cultural Gerontology, 26,* 279–298.

Moore, R. C., Eyler, L. T., Mausbach, B. T., Zlatar, Z. Z., Thompson, W. K., Peavy, G., … Jeste, D. V. (2015). Complex interplay between health and successful aging: Role of perceived stress, resilience, and social support. *The American Journal of Geriatric Psychiatry, 23*(6), 622–632.

Parker, M. G., & Thorslund, M. (2007). Health trends in the elderly population: Getting better and getting worse. *The Gerontologist, 47,* 150–158. https://doi.org/10.1093/geront/47.2.150.

Putnam, M., Tang, F., Brooks-Danso, A., Pickard, J., & Morrow-Howell, N. (2007). Professionals' beliefs about nursing home regulations in Missouri. *Journal of Applied Gerontology, 26*(3), 290–304.

Reimer, H. D., & Keller, H. H. (2009). Mealtimes in nursing homes: Striving for person-centered care. *Journal of Nutrition for the Elderly, 28*(4), 327–347. https://doi.org/10.1080/01639360903417066.

Rémillard-Boilard, S., Buffel, T., & Philipson, C. (2017). Involving older residents in age-friendly developments: From information to coproduction mechanisms. *Journal of Housing for the Elderly, 31,* 146–159. https://doi.org/10.1080/02763893.2017.1309932.

Riskin, C. (2000). *China human development report 1999: Transition and the state*. New York, NY: Oxford University Press.

Scheidt, R. J., Humpherys, D. R., & Yorgason, J. B. (1999). Successful aging: What's not to like? *Journal of Applied Gerontology, 18,* 277–282.

Shanghai Civil Affairs Bureau. (2008). Shanghai Civil Affairs Bureau's notification on encouraging to set up community-based dining sites for elders. Retrieved from http://mzj.sh.gov.cn/gb/shmzj/node8/node15/node55/node230/node278/userobject1ai11970.html [Chinese].

Shanghai Civil Affairs Bureau. (2013). Notice on publishing "Constructing Shanghai as an age-friendly city guidelines." Retrieved from http://mzj.sh.gov.cn/gb/shmzj/node8/node194/u1ai36526.html [Chinese].

Shanghai Government. (2007). Shanghai government dedicates to "90-7-3" eldercare framework. Retrieved from http://www.shanghai.gov.cn/shanghai/node2314/node2315/node4411/userobject21ai238364.html [Chinese].

Shanghai Research Center on Aging. (2008). Shanghai 2007 aging population statistics. Retrieved from http://www.shrca.org.cn/4673.html [Chinese].

Shanghai Research Center on Aging. (2018). Shanghai aging population statistics 2017. Retrieved from http://www.shrca.org.cn/5783.html [Chinese].

Shaw, B. A., Krause, N., Liang, J., & Bennett, J. (2007). Tracking changes in social relations throughout late life. *The Journals of Gerontology Series B: Psychological Sciences and Social Sciences, 62*(2), S90–S99.

Westermann, C., Kozak, A., Harling, M., & Nienhaus, A. (2014). Burnout intervention studies for inpatient elderly care nursing staff: Systematic literature review. *International Journal of Nursing Studies, 51,* 63–71. https://doi.org/10.1016/j.ijnurstu.2012.12.001.

Whyte, M. K. (2004). Filial obligations in Chinese families: Paradoxes of modernization. In C. Ikels (Ed.), *Filial piety: Practice and discourse in contemporary East Asia* (pp. 106–127). Stanford, CA: Stanford University Press.

World Health Organization. (2007). *Global aging-friendly cities: A guide*. Geneva: Switzerland.

Yu, H. (2013, July 8). When filial piety is the law. *New York Times*. Retrieved from http://www.nytimes.com/2013/07/08/opinion/yu-when-filial-piety-is-the-law.html?hp&_r=2&.

Zeng, Y., Feng, Q., Hesketh, T., Chistensen, K., & Vaupel, J. W. (2017). Survival, disabilities in activities of daily living, and physical and cognitive functioning among the old-old in China: A cohort study. *The Lancet, 389*(10079), 1619–1629.

The manufacturer's authorised representative in the EU is Springer
Nature Customer Service Centre GmbH, Europaplatz 3, 69115 Heidelberg,
Germany. If you have any concerns regarding our products, please
contact ProductSafety@springernature.com

Printed and bound by CPI Group (UK) Ltd, Croydon, CR0 4YY

30/09/2026

02225502-0002